DYNAMICS OF MARKETS
The New Financial Economics

SECOND EDITION

This second edition presents the advances made in finance market analysis since 2005. The book provides a careful introduction to stochastic methods along with approximate ensembles for a single, historic time series.

This new edition explains the history leading up to the biggest economic disaster of the 21st century. Empirical evidence for finance market instability under deregulation is given, together with a history of the explosion of the US Dollar worldwide. A model shows how bounds set by a central bank stabilized foreign exchange in the gold standard era, illustrating the effect of regulations. The book presents economic and finance theory thoroughly and critically, including rational expectations, cointegration, and ARCH/GARCH methods, and replaces several of those misconceptions with empirically based ideas.

This book will interest finance theorists, traders, economists, physicists and engineers, and leads the reader to the frontier of research in time series analysis.

JOSEPH L. MCCAULEY is Professor of Physics at the University of Houston, and is an advisory board member for the Econophysics Forum. He has contributed to statistical physics, the theory of superfluids, nonlinear dynamics, cosmology, econophysics, economics, and finance theory.

'A thought provoking book. It does not only argue convincingly that the 'King – of orthodox economic theory – is naked', but offers a challenging economic alternative interpretation regarding especially the dynamics of financial markets.'

Giovanni Dosi, Laboratory of Economics and Management,
Sant'Anna School of Advanced Studies, Pisa

'The heart of McCauley's book is a closely-reasoned critique of financial-economic mathematical modeling practice. McCauley's demonstration of the incompatibility between the assumptions of market-clearing equilibrium and informational efficiency is stunning, and sheds much-needed light on the mathematical modeling failures revealed by the financial melt-down. His unvarnished criticisms of neoclassical economic doctrine deserve equal attention. McCauley opens the windows of the self-referential world of economics to the fresh air of a mathematical physics point of view grounded in economic history and common sense. Neither monetarist, neoclassical, nor Keynesian schools of economics will take much comfort from McCauley's work, but they all have a lot to learn from it.'

Duncan K. Foley, Leo Model Professor, New School for
Social Research and External Professor, Santa Fe Institute

'McCauley's mathematically and empirically rigorous Dynamics of Markets is one of those rare works which is challenging, not only to an intellectual orthodoxy (neoclassical economics), but also to its fledgling rival (econophysics). Neoclassical economics and finance theory receive justifiably dismissive treatments for failing empirically, but some econophysics contributions also distort empirical data– notably McCauley shows that "fat tails" in data can be the result of applying an unjustified binning process to nonstationary data. McCauley's essential messages for the future of economics after the Global Financial Crisis is that "There is no statistical evidence for Adam Smith's Invisible Hand", and that the hand that does exist and must be understood is both non-stationary and far from equilibrium.'

Steve Keen, School of Economics and Finance,
University of Western Sydney

DYNAMICS OF MARKETS

The New Financial Economics

SECOND EDITION

JOSEPH L. MCCAULEY
University of Houston

CAMBRIDGE UNIVERSITY PRESS
Cambridge, New York, Melbourne, Madrid, Cape Town, Singapore, São Paulo, Delhi

Cambridge University Press
The Edinburgh Building, Cambridge CB2 8RU, UK

Published in the United States of America by Cambridge University Press, New York

www.cambridge.org
Information on this title: www.cambridge.org/9780521429627

© J. McCauley 2009

This publication is in copyright. Subject to statutory exception
and to the provisions of relevant collective licensing agreements,
no reproduction of any part may take place without
the written permission of Cambridge University Press.

First published 2004
Reprinted 2006
Paperback edition 2007
Second edition 2009

Printed in the United Kingdom at the University Press, Cambridge

A catalog record for this publication is available from the British Library

Library of Congress Cataloging-in-Publication Data
McCauley, Joseph L.
Dynamics of markets : the new financial economics / Joseph L. McCauley. – 2nd ed.
p. cm.
Includes bibliographical references.
ISBN 978-0-521-42962-7 (hardback)
1. Finance–Mathematical models. 2. Finance–Statistical methods. 3. Business mathematics.
4. Markets–Mathematical models. 5. Statistical physics. I. Title.

HG106.M4 2009
332.01'5195–dc22

2009015596

ISBN 978-0-521-42962-7 hardback

Cambridge University Press has no responsibility for the persistence or
accuracy of URLs for external or third-party Internet websites referred to
in this publication, and does not guarantee that any content on such
websites is, or will remain, accurate or appropriate.

For my stimulating partner Cornelia, who worked very hard
and effectively helping me to improve the text in both editions,
and for our sons, Finn and Hans.

Contents

Preface to the second edition   page xi

1 Econophysics: why and what   1
    1.1 Why econophysics?   1
    1.2 Invariance principles and laws of nature   4
    1.3 Humanly invented law can always be violated   5
    1.4 Origins of econophysics   7
    1.5 A new direction in econophysics   8

2 Neo-classical economic theory   10
    2.1 Why study "optimizing behavior"?   10
    2.2 Dissecting neo-classical economic theory (microeconomics)   12
    2.3 The myth of equilibrium via perfect information   18
    2.4 How many green jackets does a consumer want?   24
    2.5 Macroeconomics   25

3 Probability and stochastic processes   29
    3.1 Elementary rules of probability theory   29
    3.2 Ensemble averages formed empirically   30
    3.3 The characteristic function   32
    3.4 Transformations of random variables   33
    3.5 Laws of large numbers   34
    3.6 Examples of theoretical distributions   38
    3.7 Stochastic processes   43
    3.8 Stochastic calculus   57
    3.9 Ito processes   63
    3.10 Martingales and backward-time diffusion   77

4	**Introduction to financial economics**	**80**
	4.1 What does no-arbitrage mean?	80
	4.2 Nonfalsifiable notions of value	82
	4.3 The Gambler's Ruin	84
	4.4 The Modigliani–Miller argument	85
	4.5 Excess demand in uncertain markets	89
	4.6 Misidentification of equilibrium in economics and finance	91
	4.7 Searching for Adam Smith's Unreliable Hand	93
	4.8 Martingale markets (efficient markets)	94
	4.9 Stationary markets: value and inefficiency	98
	4.10 Black's "equilibrium": dreams of recurrence in the market	101
	4.11 Value in real, nonstationary markets	102
	4.12 Liquidity, noise traders, crashes, and fat tails	103
	4.13 Long-term capital management	105
5	**Introduction to portfolio selection theory**	**107**
	5.1 Introduction	107
	5.2 Risk and return	107
	5.3 Diversification and correlations	109
	5.4 The CAPM portfolio selection strategy	113
	5.5 Hedging with options	117
	5.6 Stock shares as options on a firm's assets	120
	5.7 The Black–Scholes model	122
	5.8 The CAPM option pricing strategy	124
	5.9 Backward-time diffusion: solving the Black–Scholes pde	127
	5.10 Enron 2002	130
6	**Scaling, pair correlations, and conditional densities**	**133**
	6.1 Hurst exponent scaling	133
	6.2 Selfsimilar Ito processes	135
	6.3 Long time increment correlations	139
	6.4 The minimal description of dynamics	145
	6.5 Scaling of correlations and conditional probabilities?	145
7	**Statistical ensembles:** *deducing dynamics from time series*	**148**
	7.1 Detrending economic variables	148
	7.2 Ensemble averages constructed from time series	149
	7.3 Time series analysis	152
	7.4 Deducing dynamics from time series	162

	7.5	Early evidence for variable diffusion models	167
	7.6	Volatility measures	167
	7.7	Spurious stylized facts	168
	7.8	An sde for increments?	173
	7.9	Topological inequivalence of stationary and nonstationary processes	173
8	**Martingale option pricing**	**176**
	8.1	Introduction	176
	8.2	Fair option pricing	178
	8.3	Pricing options approximately via the exponential density	182
	8.4	Option pricing with fat tails	185
	8.5	Portfolio insurance and the 1987 crash	186
	8.6	Collateralized mortgage obligations	186
9	**FX market globalization:** *evolution of the Dollar to worldwide reserve currency*	**188**
	9.1	Introduction	188
	9.2	The money supply and nonconservation of money	189
	9.3	The gold standard	190
	9.4	How FX market stability worked on the gold standard	190
	9.5	FX markets from WWI to WWII	194
	9.6	The era of "adjustable pegged" FX rates	196
	9.7	Emergence of deregulation	197
	9.8	Deficits, the money supply, and inflation	204
	9.9	Derivatives and shadow banking	208
	9.10	Theory of value under instability	211
	9.11	How may regulations change the market?	212
10	**Macroeconomics and econometrics:** *regression models vs empirically based modeling*	**214**
	10.1	Introduction	214
	10.2	Muth's rational expectations	216
	10.3	Rational expectations in stationary markets	219
	10.4	Toy models of monetary policy	222
	10.5	The monetarist argument against government intervention	224
	10.6	Rational expectations in a nonstationary world	225
	10.7	Integration I(d) and cointegration	226
	10.8	ARCH and GARCH models of volatility	238

11 Complexity   241
11.1 Reductionism and holism   241
11.2 What does "complex" mean?   244
11.3 Replication, mutations, and reliability   253
11.4 Emergence and self-organization   256

References   261
Index   268

Preface to the second edition

This book provides a thorough introduction to econophysics and finance market theory, and leads the reader from the basics to the frontiers of research. These are good times for econophysics with emphasis on market instability, and bad times for the standard economic theory that teaches stable equilibrium of markets. I now explain how the new volume differs in detail from the first edition.

The first edition of *Dynamics of Markets* (2004) was based largely on our discovery of diffusive dynamics of the exponential model, and more generally on the dynamics of Markovian models with variable diffusion coefficients. Since that time, the progress by the University of Houston Group (Kevin Bassler, Gemunu Gunaratne, and me) has produced a far more advanced market dynamics theory based on our initial discovery. The present book includes our discoveries since 2004. In particular, we've understood the limitations of scaling and one-point densities: given a scaling process, only the one-point density can scale, the transition density and all higher-order densities do not and cannot scale, and a one-point density (as Hänggi and Thomas pointed out over 30 years ago) cannot be used to identify an underlying stochastic process. Even pair correlations do not scale. It follows that scaling cannot be used to determine the dynamics that generated a time series. In particular, scaling is not an indication of long time correlations, and we exhibit scaling Markov models to illustrate that point. Our focus in this edition is therefore on the pair correlations and transition densities for stochastic processes, representing the minimum level of knowledge required to identify (or rule out) a class of stochastic processes.

The central advances are our 2007 foreign exchange (FX) data analysis, and the Martingale diffusion theory that it indicates. We therefore focus from the start on the pair correlations of stochastic processes needed to understand and characterize a class of stochastic processes. The form of the pair correlations tells us whether we're dealing with Martingale dynamics, or with the dynamics of long time pair correlations like fractional Brownian motion. The stochastic processes with pair correlations agreeing empirically with detrended finance data are Martingales, and the addition of drift to a Martingale yields an Ito process. We therefore emphasize Ito processes, which are diffusive processes with uncorrelated noise increments. Stated otherwise, the Martingale is the generalization of the Wiener process to processes with general (x,t)-dependent diffusion coefficients. In physics x denotes position; in finance and macroeconomics x denotes the logarithm of a price.

A much more complete development of the theory of diffusive stochastic processes is provided in this text than in the first edition, with simple examples showing how to apply Ito calculus. We show that stationary markets cannot be efficient, and vice versa, and show how money could systematically be made with little or no risk by betting in a stationary market. The Dollar on the gold standard provides the illuminating example. The efficient market hypothesis is derived as a Martingale condition from the absence of influence of the past on the future at the level of pair correlations. Because of nonstationarity, the analysis of an arbitrary time series is nontrivial. We show how to construct an approximate ensemble for a single historic time series like finance data, and then show how a class of dynamical models can be deduced from the statistical ensemble analysis. Our new FX data analysis is discussed in detail, showing that the dynamics in log returns is a Martingale after a time lag of 10 minutes in intraday trading, and we show how spurious stylized facts are generated by a common but wrong method of data analysis based on time averages.

Here are some main points from each chapter. In Chapter 1 physics is contrasted with economics, and Wigner's description of the basis in symmetry for natural law is surveyed. We point out that some sort of regularity in a time series is required if a model is to be deduced. Chapter 2 introduces neo-classical economics and its falsification by Osborne. Increments, pair correlations, and transition densities are developed as the basis for the theory of stochastic processes in Chapter 3, where enlightening and nonstandard derivations of Kolmogorov's two partial differential equations (pdes) are provided. Chapter 4 provides a solid basis for much in the rest of the text. Therein, we explain both stationary and efficient markets and show how one excludes the other, and generalize the neo-classical notion of "value" to uncertain markets. The efficient market hypothesis is derived from the assumptions that past returns are uncorrelated with future returns increments, and an error in Fama's discussion is corrected. Standard misconceptions about market equilibrium and stability are exposed and dispelled. Chapter 5

covers standard topics like the Capital Asset Pricing Model and the original
Black–Scholes model. Chapter 6 covers scaling processes and also fractional
Brownian motion in detail, and shows why transition densities and pair
correlations cannot scale for a "scaling process," relegating scaling to no
interest when the aim is to identify the dynamics from a time series. Statistical
ensembles and their basis in vanishing correlations of initial data are presented
in fairly complete detail in Chapter 7. An approximate statistical ensemble is
constructed for the analysis of a single, historic time series, where it's shown
that certain averages can be reliably measured, others not. Regularities in
traders' daily behavior are reflected in the time variation of the ensemble
average diffusion coefficient of the Martingale describing the finance market.
I also show how and why standard time averages ("sliding windows") on
nonstationary time series cannot be expected to converge to any limit in
probability. I use our FX analysis to illustrate the basis for pinning down
classes of mathematical models in the social sciences and beyond. I then show
how spurious stylized facts like fat tails and misleading Hurst exponents are
generated when time averages are used on nonstationary time series. Volatility
is introduced and discussed (and is discussed in detail in Chapter 10). In
Chapter 8 I provide a basic introduction to generalized Black–Scholes option
pricing for arbitrary Ito processes, and show that for arbitrary drift and
diffusion coefficients the generalized Black–Scholes pde yields Martingale
option prices. We discuss how invalid liquidity assumptions can lead to
market crashes, and begin to discuss the derivatives-based credit bubble that
burst in September, 2008. Chapter 9 presents the history of the Dollar and FX
since the gold standard as the prime example of the instability generated by
deregulation, and ends with a discussion of the worldwide financial crisis and
the money supply. The main point, illustrated by the Dollar on and off the
gold standard, is that markets without strong regulations can be expected to
show instability. The notion that deregulation and free trade are maximally
beneficial to society is a neo-classical assumption with nations taken as agents.
We discuss the mortgage credit bubble and shadow banking, and why credit
creation has exploded worldwide via derivatives. Chapter 10 presents standard econometric methods of regression analysis in macroeconomic theory,
based on the untenable assumption of market stability. Cointegration and
integration $I(d)$ are presented, and the inapplicability of those assumptions to
real data are discussed. I show that the Lucas policy critique is based on a
severely restricted and nonempirically based monetary model, and explain via
counterexamples that nonstationarity in empirical data cannot be eliminated
by cointegration in regression analysis. ARCH and GARCH regression
models are shown to violate observed Martingale finance markets. The final

chapter on complexity is an enlargement of the original one, and includes the idea of emergence (in biology).

In short, we offer an alternative to the standard macroeconomic theory, which is based on overly restrictive regression models, and we name this alternative "The New Financial Economics." This is one step toward the goal stated in *Nature* (Ball, 2006), that econophysics will eventually replace micro- and macroeconomic theory in both the classrooms and the boardrooms. This second edition appears at the right time, when ordinary people (if not academic theorists) are questioning the use of ad hoc models as the basis for finance trading, and are questioning the assumption that unregulated markets provide the best of all possible worlds as their jobs are transferred eastward to cheaper, unorganized labor. The lessons of the local labor battles in the west from the early part of the twentieth century, where unions had to be established so that workers could gain a living wage from the owners of capital, have been lost. With the fall of the Berlin Wall in 1989, and then the USSR in 1991, it was largely assumed that *laissez faire* had triumphed as regulated Europe began to follow Reagan-Thatcher-Friedman policies and deregulate, but the failed promise of Pareto optimality of the *laissez faire* program has now been exposed by the popping of the worldwide credit bubble. Deregulation has helped the east, and has hurt the west. The big question is how ordinary workers will make a living in the future. Such questions are not discussed in financial engineering classes. The student who wants to learn financial engineering is advised to put away this book, which focuses on understanding markets rather than on making ad hoc models to sell to well-heeled buyers, and instead to consult one of the many fine financial math books available (e.g. Baxter and Rennie, 1995).

This book can be studied as follows. First, for the mathematically challenged reader, Chapters 1, 2, 4, and 9 can be read while ignoring the math. In Chapters 1, 3, 4, 6, 7, and 9 the math and main ideas are fully developed. Chapter 7 is the high point, but Chapter 9 broadens the perspective from FX markets to the role of the money supply in international trade and finance. Chapters 1–5 provide a basic introduction to elementary ideas of finance combined with the math. The original Black–Scholes model in Chapter 5 can be understood by restricting the math in Chapter 3 to basic Ito calculus and the Fokker–Planck equation. Chapter 10 requires Chapters 3 and 7 as background, and is further illuminated by the analysis of Chapter 9. The one-semester econophysics course at the University of Houston consists regularly of Chapters 1 and 2 (lightly covered), Chapters 4–8 (heavily covered). Chapter 9 (which began as an invited talk for the 2007 Geilo NATO-ASI) was included once. Chapter 10 was developed later. Chapter 11 presents my understanding of complexity in dynamics.

Chapter 3 is quite long because it's unusually complete; topics are included there that are either hard or impossible to find in other texts. Through Chapter 5, the following parts can be ignored: 3.5–3.53, 3.6.8–9, 3.7.4, 3.9. For Chapter 6 one needs part 3.6.9. For Chapter 7 one needs parts 3.5–3.5.3 and 3.6.8. Chapter 8 is based on Section 3.9.

I'm extremely grateful for key discussions and criticism (mainly via email) to Harry Thomas, Enrico Scalas, Giulio Bottazzi, Søren Johansen, Giovanni Dosi, Duncan Foley, Peter R. Hansen, Steve Keen, Jonathan Batten, and Barkley Rosser. I'm also grateful to Doyne Farmer, Giulia Rotundo, Emanuel Derman, Peter Toke Heden Algren, and Bernard Meister for (largely email) discussions. My friend Vela Velupillai has encouraged and supported my work strongly, even to the extent of having made me a Fellow in Economics at the National University of Ireland, Galway, before his health forced him to give up his position as the John E. Cairnes Professor there. Useful conversations with Stefano Zambelli, Mauro Gallegatti, Sorin Solomon, David Bree, Simona Cantono, Filipo Petroni and Roberto Tamborini are also acknowledged. My wife, hiking partner, and local editor, Cornelia Küffner, critically read the entire manuscript (skipping the math) and made useful suggestions for a better presentation. Finally, I'm grateful to Simon Capelin for the opportunity to publish this revised second edition at an extremely interesting – because troublesome – time in international finance, and to Lindsay Barnes for riding herd on the project once it started.

1
Econophysics: why and what

1.1 Why econophysics?

This is the era of growing financial instability, a new era of worldwide privatization and deregulation made possible by a vast credit expansion based on the Dollar as the worldwide default reserve currency. Derivatives are unregulated and are used as a form of money creation totally beyond the control of any central bank. Standard economic theory completely rules out the possibility of such instability.

Before WWII, the expansion of a currency and consequent inflation was not possible with the Dollar regulated by gold at $35/oz. The gold standard was finally and completely abandoned by the USA in 1971 after "Eurodollars" became on the order of magnitude of the US gold supply. On the gold standard, hedging foreign currency bets apparently was not necessary. We can date our present era of inflation, credit, and high level of consumption with increasing finance market instability from the deregulation of the Dollar in 1971, and it's not accidental that both the Black–Scholes derivatives model and the legalization of large-scale options trading both date from 1973. We can contrast this reality, described in popular books by Stiglitz (2002), Morris (2008), and Soros (2008), with the teaching of equilibrium in standard academic economics texts.

Economists teach market equilibrium as the benchmark in the classroom, even while the real world of economics outside the classroom experiences no stability. There is an implicit assumption in those texts that unregulated markets are stable, as if completely free markets should somehow self-organize in a stable way.

Standard microeconomic theory is based on a deterministic equilibrium model, called neo-classical economics (Chapter 2), where perfect knowledge of the infinite future is assumed on the part of all players. That an equilibrium exists *mathematically* under totally unrealistic conditions has been proven, but that the hypothetical equilibrium is stable (or computable) or has anything at all to do with reality was never demonstrated. The generalization of the neo-classical model to uncertain but still hypothetically stable markets assumes a stationary stochastic process, and is called "rational expectations". Standard macroeconomics is based on the assumption of stationary and therefore stable economic variables. Rational expectations emerged as the dominant economic philosophy parallel to deregulation in the 1970s and 1980s, with regression analysis as the tool of choice for modeling. Regression analysis is based on the assumption of stationary noise, *but there is no solid empirical evidence for stationarity of any kind in any known market*. The only scientific alternative is to approach markets as a physicist, and ask the market data what are the underlying unstable dynamics.

Having stated our view of standard economics and our offered alternative, we now survey the historic viewpoint of physics. In particular, Galileo did not merely discover a mathematical model of nature, he discovered two inviolable local laws of nature: the law of inertia and the local law of gravity. Both of those local laws survived the Einsteinian and quantum revolutions. Following the lessons of Galileo, Kepler, and Newton, scientists have amassed indisputable evidence that mindless nature behaves mathematically lawfully. But "motion" guided by minds is an entirely different notion. Social behavior is generally complicated, it may be artificially regulated by the enforcement of human law, or it may be completely lawless. Neo-classical economists try to model human preferences using *a priori* models of behavior (utility maximization) that have been falsified. More recent work in both econophysics and economics uses agent-based modeling, which is like trying to replace thinking, hopeful, and fearful agents with fixed rules obeyed by spins on a lattice. In this text we will instead adopt an inherently macroeconomic, or phenomenological, viewpoint. We will not try to model what agents prefer or do, but instead will simply ask real markets what the observed statistics can teach us. In particular, we will try to discover regularities in the form of equations of motion for log returns of prices. The discovery of a correct class of dynamic models is far beyond the reach of regression analysis in econometrics.

The history of physics shows that mathematical law cannot be discovered from empirical data unless *something* is repeated systematically. Wigner has explained the basis for the discovery of mathematical laws of motion in local invariance principles. But the method of the natural sciences cannot be found in standard economic theorizing and data analysis. In financial economics, where no correct dynamical model has been discovered, the term "stylized facts" appears. "Stylized facts" are supposed to be certain statistical features of the data. But even there, certain hidden assumptions in statistical analysis have implicitly and unquestionably been taken for granted without checking for their validity. We'll show (Chapter 7) how a common method of data analysis leads to spurious stylized facts, to features "deduced statistically" that are really not present in the empirical data. We avoid generating spurious statistical results by constructing an approximate statistical ensemble for the analysis of a single, historic nonstationary time series.

Karl Popper only put into words what physicists since Galileo, Kepler, and Newton have done. Science consists of falsifiable propositions and theories. Falsifiable models have no free parameters to tweak that would make a wrong model fit adequate data (data with enough points for "good statistics"). A falsifiable model is specified completely by empirically measurable parameters so that, if the model is wrong, then it can be proven wrong via measurement. Examples of falsifiable models in economics and finance are neo-classical economics and the original Black–Scholes Gaussian returns model. Both models have been falsified. In science the skeptics, not the believers, must be convinced via systematic, repeatable measurements. The application of the idea of "systematic repeated observations," the notion of a statistical ensemble, is applied to the analysis of a single, historic time series in Chapter 7. The basis for the statistical ensemble is an observed repetitiveness in traders' behavior on a daily time scale. We predict a new class of falsifiable dynamical model.

In Chapter 3 we will emphasize the distinction between local and global predictions. "Local" means in a small region near a given point (x,t), whereas "global" means over large displacements $x(t,T) = x(t + T) - x(t)$ for different initial times t and large time lags T. The limitations on global predictability in perfectly well-defined deterministic dynamical systems are well defined, and inform the way that I understand and present stochastic dynamics and market models. We will distinguish local from global solutions of stochastic processes. In particular, we see no good reason to expect universality of market dynamics, and find no statistical evidence for that notion. Our analysis shows that finance markets vary in detail from one financial center to another (e.g. New York to Tokyo), and may not obey exactly the same dynamics.

The reader is encouraged to study Wigner's (1960) essay on the unreasonable effectiveness of mathematics in nature and his book *Symmetries and Reflections* (1967), and Velupillai's corresponding essay on the unreasonable ineffectiveness of mathematics in economics (2005). We turn next to Wigner's explanation of the basis for discovering laws of motion: local invariance principles.

1.2 Invariance principles and laws of nature

It's important to have a clear picture of just how and why standard economic theorizing differs from theoretical physics. To see the difference, the reader may compare any micro- or macroeconomics text with any elementary physics or astronomy text. The former describes only mental constructs like equilibrium of supply and demand that are not observed in real markets; the latter present the accurate mathematical descriptions of the historic experiments and observations on which physics and astronomy are based. In particular, where equilibrium is discussed, real examples are presented (a flower pot hanging from a ceiling, for example). Physics and astronomy are about the known mathematical laws of nature. Economics texts are about stable equilibria that do not exist in any known market. Why, in contrast, has mathematics worked so precisely in the description of nature?

Eugene Wigner, one of the greatest physicists of the twentieth century and the acknowledged expert in symmetry principles, wrote most clearly about the question: why are we able to discover mathematical laws of nature? (Wigner, 1967) An historic example points to the answer. In order to combat the prevailing Aristotelian ideas, Galileo proposed an experiment to show that relative motion doesn't matter. Motivated by the Copernican idea, his aim was to explain why, if the earth moves, we don't feel the motion. His proposed experiment: drop a ball from the mast of a uniformly moving ship on a smooth sea. It will, he asserted, fall parallel to the mast just as if the ship were at rest. Galileo's starting point for discovering physics was therefore the principle of relativity. Galileo's famous thought experiment would have made no sense were the earth not a local inertial frame for times on the order of seconds or minutes.[1] Nor would it have made sense if initial conditions like absolute position and absolute time mattered.

The known mathematical laws of nature, the laws of physics, do not change on any observable time scale. Physicists and chemists were able to discover that nature obeys inviolable mathematical laws only because those laws are grounded in local invariance principles, local invariance with respect to frames moving at constant velocity (principle of relativity), local translational invariance, local rotational invariance and local time-translational invariance. These local invariances are the same whether we discuss Newtonian mechanics, general relativity, or quantum mechanics. Were it not for these underlying invariance principles it would have been impossible to discover

---

[1] There exist in the universe only local inertial frames, those locally in free fall in the net gravitaional field of other bodies; there are no global inertial frames as Mach and Newton assumed. See Barbour (1998) for a fascinating and detailed account of the history of mechanics.

mathematical laws of nature in the first place. Why is this? Because the local invariances form the theoretical basis for *repeatable identical experiments/observations* whose results can be reproduced by different observers independently of where and at what time the observations are made, and independently of the state of relative motion of the observational machinery. This leads us to the idea of a statistical ensemble based on repetition, a main topic of Chapter 7.

In physics, astronomy, and chemistry, we do not have merely *models* of the behavior of matter. Instead, we know mathematical laws of nature that cannot be violated intentionally. They are beyond the possibility of human invention, intervention, or convention, as Alan Turing, the father of modern computability theory, said of arithmetic in his famous paper defining computability. Our discussion above informs us that *something* must be systematically repeated if we're to have any chance to discover equations of motion. The motion of the ball is trivial periodic; it has a cycle of period zero. A simple pendulum has a cycle of period one. Finance data don't generate deterministic cycles, but instead, as we'll show, exhibit a certain statistical periodicity.

Mathematical laws of nature have been established by repeatable identical (to within some decimal precision) experiments or observations. Our aim is to try to mimic this so far as is possible in finance. To qualify as science, a model must be falsifiable. A falsifiable theory or model is one with few enough parameters and definite enough predictions, preferably of some new phenomenon, that it can be tested observationally and, if wrong, can be proven wrong. A theory is not established because its promoters believe it. To gain wide acceptance, a theory must convince the skeptics, who should perform their own experiments or observations. In economics this has not been the method of choice. As various books and articles have correctly observed, textbook economic theory is not empirically based but rather is an example of socially constructed modeling. Rational expectations (Chapter 10) provides the latest example.

1.3 Humanly invented law can always be violated

Physics and economics are completely different in nature. In economics, in contrast with physics, there exist no known inviolable mathematical laws of "motion"/behavior. Instead, economic law is either legislated law, dictatorial edict, contract, or in tribal societies the rule of tradition. Economic "law," like any legislated law or social contract, can always be violated by willful people and groups. The idea of falsification via observation has not yet taken root in adequately thick topsoil. Instead, an internal logic system called neo-classical economic theory was invented via postulation and still dominates academic economics, the last contributor being Robert Lucas, who's given credit for the "rational expectations revolution" in economic theory. Neo-classical economics is not derived from empirical data. The good news is that the general predictions of the theory are specific and have been falsified. The bad news is that this is still the standard theory taught in economics textbooks, where there are many "graphs" but few if any that can be obtained from or justified by unmassaged, real market data.

In his very readable book *Intermediate Microeconomics*, Hal Varian (1999), who was a dynamical systems theorist before he was an economist, writes that much of (neo-classical) economics (theory) is based on two principles:

The optimization principle. People try to choose the best patterns of consumption they can afford.

The equilibrium principle. Prices adjust until the amount that people demand of something is equal to the amount that is supplied.

Both of these principles sound like common sense, and we will see that they turn out to be more akin to common sense than to science. They have been postulated as describing markets, but lack the required empirical underpinning.

Because the laws of physics, or better said the known laws of nature, are based on local invariance principles, they are independent of initial conditions like absolute time, absolute position in the universe, and absolute orientation. We cannot say the same about markets: socio-economic behavior is not necessarily universal but may vary from country to country. Mexico is not necessarily like China, which is certainly not like the USA or Germany. Many econophysicists, in agreement with economists, would like to ignore the details and hope that a single universal "law of motion" governs markets, but that idea remains only a hope. We will see in Chapter 4 that there is but a single known law of socio-economic invariance, and that is not enough for universally valid market dynamics.

The best we can reasonably hope for in economic theory is a model that captures and reproduces the essentials of historical data for specific markets during some epoch, like finance markets since c. 1990. We can try to describe mathematically what has happened in the past, but there is no guarantee that the future will be the same. Insurance companies provide an example. There, historic statistics are used with success in making money under normally expected circumstances, but occasionally there comes a "surprise" whose risk was not estimated correctly based on past statistics, and the companies consequently lose a lot of money through paying unexpected claims.

Some people may fail to see that there is a difference between economics and the hardest unsolved problems in physics. One might object: we can't solve the Navier-Stokes equations for turbulence because of the butterfly effect or the computational complexity of the solutions of those equations, so what's the difference with economics? Economics cannot be fairly compared with turbulence. In fluid mechanics we know the equations of motion based on Galilean invariance principles. In turbulence theory we cannot predict the weather. However, we *understand* the weather physically and can describe it qualitatively and reliably based on the equations of thermo-hydrodynamics. We understand very well the physics of formation and motion of hurricanes and tornadoes, even if we cannot predict when and where they will hit. No comparable basis for qualitative understanding exists in economic theory.

1.4 Origins of econophysics

Clearly, econophysics should not try to imitate academic economic theory, nor should econophysics rely on standard econometric methods. We are not trying to make incremental improvements in theory, as Yi-Cheng Zhang has so poetically put it, we're trying instead to replace the standard models and methods with entirely new results. Econophysics began in this spirit in 1958 with M. F. M. Osborne's discovery of Gaussian stock market returns (the lognormal pricing model), Mandelbrot's emphasis on Martingales for describing hard-to-beat markets, and then Osborne's falsification in 1977 of the supply–demand curves. From the practical side, a supply–demand mismatch of physics PhDs to academic jobs, and new research opportunities in practical finance, drew many physicists to "Wall Street." Physics funding had exploded in America after Sputnik was launched by the USSR in October, 1957, but had tapered off by 1971, when academic jobs in physics began to dry up (see Derman's informative autobiography (2004), which is a history of that era). In 1973 the Black–Scholes theory of option pricing was finally published after a struggle of several years against editors who insisted that finance wasn't economics, and large-scale options trading was legalized at the same time. The advent of deregulation as a dominant government philosophy in the 1980s (along with the opening of China to investment c. 1980, following the Nixon-Kissinger visit to Chairman Mao and Chou En-Lai in 1973), the collapse of the USSR in 1989–1991, and the explosion of computing technology in the 1980s all played determining roles in the globalization of capital. With computerization, finance data became more accurate and more reliable than fluid turbulence data, inviting physicists to build falsifiable finance models. All of these developments opened the door to the globalization of trade and capital and led to a demand on modeling and data analysis in finance that many physicists have found to be either interesting or lucrative.

1.5 A new direction in econophysics

One can ask why physicists believe that they're more qualified than economists to explain economic phenomena, and *if* physicists *then* why not also mathematicians, chemists, and biologists? Mathematicians dominate both theoretical economics and financial engineering, and by training and culture they are a strongly postulatory tribe that at worst ignores real market data, and at best (financial engineering) proves powerful theorems about Gaussian models while introducing no new empirically based models to solve the fundamental problem of market dynamics (see for example the closing words in Steele's (2000) book!). Chemists and biologists are certainly empirically oriented, but are trained to focus on details that physicists usually find boring. Physicists are trained to see the connections between seemingly different phenomena, to try to get a glimpse of the big picture, and to present the simplest possible mathematical description of a phenomenon that includes no more factors than are necessary to describe the empirical data. Physicists are trained to isolate cause and effect. A good physicist like Feynman has more in common with a radio or car repairman than with a mathematician. A few highlights of a debate between econophysicists and economists can be found in Gallegati *et al.* (2006), Ball (2006), and McCauley (2006). An interesting discussion of an entirely different nature can be found in Solomon and Levy (2003).

Since the word was coined by Gene Stanley in 1995 (Mantegna and Stanley, 1999), the term econophysics has been characterized largely by three main directions, not necessarily mutually exclusive. First, there was the thorough mathematical solution of the Minority Game inspired by the Fribourg school of econophysics (Challet *et al.*, 2005), and related models of agent-based trading (Maslov, 2000). That work partly evolved later into studies of networks (Caldarelli, 2007) and "reputation systems" (Masum and Zhang, 2004). The foray into finance is illustrated by Dacorogna *et al.* (2001), Farmer (1999), and Bouchaud and Potters (2000). Models of market crashes have been constructed by A. Johansen and Sornette (2000). Most popular, however, has been the reliance on econophysics as the attempt to explain economic and finance data by scaling laws (the Hurst exponent) and fat-tailed probability distributions. The work on fat tails was initiated historically by Pareto and was revived by Mandelbrot around 1960. Since 1995, fat tails and scaling studies have been inspired by the Boston School led by Gene Stanley, who also opened Physica A to econophysics. Econophysics is still unrecognized as science by the American Physical Society, but fortunately the European Physical Society has had a Finance and Physics section since 1999 or earlier. Without Gene Stanley and Physics A, econophysics would never have gotten off the ground. Hurst exponent scaling was also emphasized in the earlier era by Mandelbrot, with his papers on rescaled range (R/S) analysis and fractional Brownian motion. If we would judge what econophysics is by the number of papers in the field, we would say that the main ideas of econophysics are agent-based models, fat tails, and scaling. But this is not enough to determine the underlying market dynamics.

Blazing a new trail, we offer an alternative approach to econophysics. We follow Osborne's lead (and validate Mandelbrot's Martingale efficient market hypothesis) and focus on the discovery of falsifiable classes of market dynamics models deduced directly from empirical data. In particular, we will present evidence for diffusive models that don't scale in log returns, nor do we find evidence for fat tails in log returns. We offer a view of finance market dynamics that contradicts the standard so-called stylized facts. Our method of analysis, unlike the other approaches, is based on statistical ensembles. In particular, we do not use time averages ("sliding windows") on nonstationary time series.

Econophysics does not mean lifting tools and models from statistical physics and then applying them directly to economics. Economics is not like chemistry, where all results follow at least in principle from physics. Neither is economics a trivial science that can be formulated and solved by transferring methods and ideas directly from physics, mathematics, or from any other field. We use the theory of stochastic processes both in data analysis and modeling, but we've had to invent new classes of stochastic models, and have found it necessary to clarify some older mathematical ideas, in order to understand finance markets. As Lars Onsager once asserted, a theoretical physicist should not start with a mathematical tool and then look around for data to explain. Instead, a "real theorist" should study the data and invent the required mathematical tools. That's what Galileo, Kepler, and Newton did. That's also what Lars did when he solved the 2D Ising model, and also earlier when he produced an exact solution to the pdes describing the dissociation and recombination of ions of a weak electrolyte in an electric field. Both were amazing mathematical feats, and the latter was directly applicable to experimental data. Econophysics, simply stated, means following the example of physics in observing and modeling markets.

2
Neo-classical economic theory

2.1 Why study "optimizing behavior"?

Globalization via deregulation and privatization is supported by the implicit and widespread belief in an economic model that teaches the avoidance of government intervention in socio-economic life. The old *laissez faire* belief was revived in the Reagan-Thatcher era, and then gained ground explosively after the collapse of central planning in communist countries. The old fight through the 1970s was between the idea of regulated markets in the west and strong central planning under communism. The question for our era is whether markets should be regulated for social purposes, as they were in western Europe prior to the fall of the wall,[1] or whether the current *laissez faire* binge will continue in spite of its inherent financial instabilities and the irreversible loss of jobs in previously well-off western nations. In particular, *laissez faire* teaches that regulations should have been avoided, and this has led to the peculiar problem that financial derivatives are a highly leveraged and unregulated form of credit creation. In contrast, the standard economic theory to be described in this chapter does not admit "money" in any form, shape, or fashion.

The "losing side" in the Cold War has adopted capitalism with a vengeance, and is now beating its former enemies: China and Russia, as of 2007, sit on the largest Dollar reserves in the world. With imports outrunning exports in the west, the problems that follow from deregulation and privatization are now felt in the so-called "First World" countries: degradation of the currency, so far mainly the Dollar, and unemployment due to the systematic loss of manufacturing capacity to cheap labor. The financial pressure to deregulate everything

---

[1] The vast middle ground represented by the regulation of free markets, along with the idea that markets do not necessarily provide the best solution to all social problems, is not taught by "Pareto efficiency" in the standard neo-classical model.

and let the cards fall as they may has reigned effectively unchallenged from 1981 until the effects of the 2007 derivatives-based credit bubble started to deflate. As Keynes once wrote, "practical men, who believe themselves to be quite exempt from any intellectual influences, are usually the slaves of some defunct economist." What are the defunct ideas that practical men like bankers, bureaucrats, and politicians operate under today?

The dominant theoretical economic underpinning for the unregulated free market is provided by neo-classical equilibrium theory, also called optimizing behavior, and is taught in standard economics texts. As Morris (2008) has written, the marriage of (neo-classical) economics with high-powered mathematics has led to the illusion that economics has become a science. The most recent form of the basic theoretical assumptions has been advanced in macroeconomics under the heading "rational expectations," which we will cover in detail in Chapter 10. The most basic assumptions are the same there as in neo-classical economics: (i) optimizing behavior and (ii) implicitly stable market equilibrium. In this chapter we will explain the predictions of those basic assumptions and compare them with reality, following Osborne (1977). We will see, among other things, that although the model is used to advise governments, businesses, and international lending agencies on financial matters, the deterministic neo-classical model (microeconomics and its extrapolation to macroeconomics) relies on presumptions of stability and equilibrium in a way that completely excludes the possibility of discussing money/capital and financial markets. It is even more strange that the standard equilibrium model completely excludes the profit motive as well in describing markets: the accumulation of capital is not allowed within the confines of that model, and, because of the severe nature of the assumptions required to guarantee equilibrium, cannot be included perturbatively either. The contradiction with real markets when the neo-classical assumptions are relaxed to include stationary stochastic markets ("rational expectations") is described in Chapters 4, 9 and 10.

Economists distinguish between classical and neo-classical economic ideas. Classical theory began with Adam Smith; neo-classical with Walras, Pareto, I. Fisher, and others. Adam Smith (2000) observed society qualitatively and invented the notion of an Invisible Hand that hypothetically should match supply to demand in free markets. When politicians, businessmen, and economists assert that "I believe in the law of supply and demand" they implicitly assume that Smith's Invisible Hand is in firm control of the market. Mathematically formulated, the Invisible Hand represents the implicit assumption that a stable equilibrium point determines market dynamics, whatever those dynamics may be. This philosophy has led to an elevated notion of the role of markets in our society. Exactly how the Invisible Hand should accomplish the self-regulation of free markets and avoid social chaos is something that economists have not been able to explain satisfactorily.

Adam Smith was not blindly against the idea of government intervention, and noted that it is sometimes necessary. He didn't assert that free markets are always the best solution to all socio-economic problems. Smith lived in a Calvinist society and also wrote a book about morals. He assumed that economic agents (consumers, producers, traders, bankers, CEOs, accountants) would exercise self-restraint so that markets would not be dominated by greed and criminality. He believed that people would regulate themselves, that self-discipline would prevent foolishness and greed from playing the dominant role in the market. This is quite different from the prevailing belief that elevates self-interest and deregulation to the level of guiding principles. Varian (1992), in his text *Intermediate Economics,* shows via a rent control example how to use neo-classical reasoning to "prove" mathematically that free market solutions are best, that any other solution is less efficient. This is the theory that students of economics are most often taught. We therefore present and discuss it critically in the next sections. In the chapter on the history of foreign exchange, we will provide examples of why unregulated society is unstable society. Interestingly enough, Adam Smith's friend David Hume introduced the equilibrium theory of foreign exchange on the gold standard in his discussion of how international trade imbalances could be remedied (Chapter 9).

Supra-governmental organizations like the World Bank and the International Monetary Fund (IMF) rely on the neo-classical equilibrium model in formulating guidelines for extending loans (Stiglitz, 2002). After you understand this chapter, you should be in a better position to understand what are the unstated ideas hidden underneath the surface whenever one of those organizations announces that a country is in violation of its rules.

2.2 Dissecting neo-classical economic theory (microeconomics)

In economic theory we speak of "agents." In neo-classical theory agents consist of consumers and producers. Let $x = (x_1, \ldots, x_n)$, where x_k denotes the quantity of asset k held or desired by a consumer. The quantity x_1 may be the number of VW Golfs, x_2 the number of Philips TV sets, x_3 the number of ice cream cones, etc. These are demanded by a consumer at prices given by $p = (p_1, \ldots, p_n)$. Neo-classical theory describes the behavior of a so-called "rational agent." By "rational agent" the neo-classicals mean the following: each consumer is assumed to perform "optimizing behavior." This means that the consumer's implicit mental calculations are assumed equivalent to maximizing a utility function $U(x)$ that is supposed to describe his ordering of preferences for these assets, limited only by his or her budget constraint M, where

$$M = \sum_{k=1}^{n} p_k x_k = \tilde{p} x \qquad (2.1)$$

Here, e.g., M equals five TV sets, each demanded at price 230 Euros, plus three VW Golfs, each wanted at 17 000 Euros, and other items. In other words, M is the sum of the number of each item wanted by the consumer times the price he or she is willing to pay for it.

That is, complex calculations and educated guesses that might require extensive information gathering, processing and interpretation capability by an agent are vastly oversimplified in this theory and are replaced instead by maximizing a simple utility function in the standard theory.

A functional form of the utility $U(x)$ cannot be deduced empirically, but U is assumed to be a concave function of x in order to model the expectation of "decreasing returns" (see Arthur (1994) for examples and models of increasing returns and feedback effects in markets). By decreasing returns, we mean that we are willing to pay less for n Ford Mondeos than we are for $n-1$, less for $n-1$ than for $n-2$, and so on. An example of such a utility is $U(x) = \ln x$ (see Figure 2.1) But what about producers?

Optimizing behavior on the part of a producer means that the producer maximizes profits subject to his or her budget constraint. We intentionally leave out savings because there is no demand for liquidity (money as cash) in this theory. The only role played here by money is as a bookkeeping device. This is explained below.

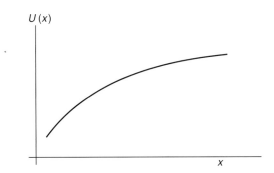

Figure 2.1 Utility vs quantity x demanded for decreasing returns.

Each consumer is supposed to maximize his or her own utility function while each producer is assumed to maximize his or her profit. As consumers we therefore maximize utility $U(x)$ subject to the budget constraint (2.1),

$$dU - \tilde{p}dx/\lambda = 0 \qquad (2.2)$$

where $1/\lambda$ is a Lagrange multiplier. We can just as well take p/λ as price p since λ only changes the price scale. This yields the following result for a consumer's demand curve, describing algebraically what the consumer is willing to pay for more and more of the same item,

$$p = \nabla U(x) = f(x) \qquad (2.3)$$

with slope p of the bidder's price decreasing toward zero as x goes to infinity, as with $U(x) = \ln x$ and $p = 1/x$, for example (see Figure 2.2). Equation (2.3) is a key prediction of neo-classical economic theory because it turns out to be falsifiable.

Some agents buy while others sell, so we must invent a corresponding supply schedule. Let $p = g(x)$ denote the asking price of assets x supplied. Common sense suggests that asking price should increase as the quantity x supplied increases (because increasing price will induce suppliers to increase production), so that neo-classical supply curves slope upward. The missing piece, so far, is that market clearing is assumed: everyone who wants to trade finds someone on the opposite side and matches up with him or her. The market clearing price is the equilibrium price, the price where total demand equals total supply. There is no dissatisfaction in such a world, dissatisfaction being quantified as excess demand, which vanishes.

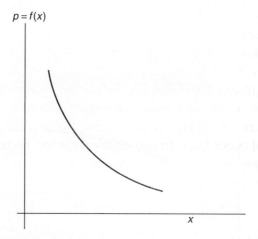

Figure 2.2 Neo-classical demand curve, downward sloping for case of decreasing returns.

2.2 Dissecting neo-classical economic theory

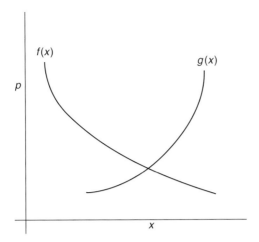

Figure 2.3 Neo-classical predictions for demand and supply curves $p = f(x)$ and $p = g(x)$ respectively. The intersection determines the idea of neo-classical equilibrium, but such equilibria are typically ruled out by the dynamics.

But even an idealized market will not start from an equilibrium point, because arbitrary initial bid and ask prices will not coincide. How, in principle, can an idealized market of utility maximizers clear itself dynamically? That is, how can a nonequilibrium market evolve toward equilibrium? To perform "optimizing behavior" the agents must know each other's demand and supply schedules (or else submit them to a central planning authority[2]) and then agree to adjust their prices to produce clearing. In this hypothetical picture everyone who wants to trade does so successfully, and this defines the equilibrium price (market clearing price), the point where the supply and demand curves $p = g(x)$ and $p = f(x)$ intersect (Figure 2.3).

There are several severe problems with this picture, and here is one: Kenneth Arrow has pointed out that supply and demand schedules for the infinite future must be presented and read by every agent (or a central market maker). Each agent must know at the initial time precisely what he or she wants for the rest of his or her life, and must allocate his or her budget accordingly. Otherwise, dissatisfaction leading to new further trades (non-equilibrium) could occur later. In neo-classical theory, no trades are made at any nonequilibrium price. Agents must exchange information, adjust their prices until equilibrium is reached, and then goods are exchanged.

---

[2] Mirowski (2002) points out that socialists were earlier interested in the theory because, if the Invisible Hand would work purely mechanically then it would mean that the market should be amenable to central planning. The idea was to simulate the free market via mechanized optimal planning rules that mimic a perfect market, and thereby beat the performance of real markets.

The vanishing of excess demand, the condition for equilibrium, can be formulated as follows: let $x_D = D(p)$ denote the quantity demanded, the demand function. Formally, this should be the inverse of $p = f(x)$ if the inverse f of D exists. Also, let $x_S = S(p)$ (the inverse of $p = g(x)$, if this inverse exists) denote the quantity supplied. In equilibrium we would have vanishing excess demand

$$x_D - x_S = D(p) - S(p) = 0 \qquad (2.4)$$

The equilibrium price, if one or more exists, solves this set of n simultaneous nonlinear equations. The excess demand is simply

$$\varepsilon(p) = D(p) - S(p) \qquad (2.5)$$

and fails to vanish away from equilibrium. Market efficiency e can be defined as

$$\varepsilon(p) = \min\left(\frac{S}{D}, \frac{D}{S}\right) \qquad (2.6)$$

so that $e = 1$ in equilibrium. Note that, more generally, efficiency e must depend on both bid and ask prices if the spread between them is large. Market clearing is equivalent to assuming 100% efficiency. One may rightly have doubts that 100% efficiency is possible in any process that depends on the gathering, exchange, and understanding of information, the production and distribution of goods and services, and other human behavior. This leads to the question whether market equilibrium can provide a good zeroth-order approximation to any real market. A good zeroth-order approximation is one where a real market can then be described accurately perturbatively, by including corrections to equilibrium as higher-order effects. That is, the equilibrium point must be stable.

A quick glance at any standard economics text (see Mankiw (2000) or Varian (1999), for example) will show that equilibrium is assumed both to exist and to be stable. The assumption of a stable equilibrium point is equivalent to assuming the existence of Adam Smith's Invisible Hand. The assumption of uniqueness, of a single global equilibrium, is equivalent to assuming the universality of the action of the Invisible Hand independently of initial conditions. Here, equilibrium would have to be an attractive fixed point with infinite basin of attraction in price space.

Arrow (Arrow and Hurwicz, 1958) and other major contributors to neo-classical economic theory went on to formulate "General Equilibrium Theory" using

$$\frac{dp}{dt} = \varepsilon(p) \qquad (2.7)$$

and discovered the mathematical conditions that guarantee a unique, stable equilibrium (again, no trades are made in the theory so long as $dp/dt \neq 0$). The equation simply assumes that prices do not change in equilibrium (where excess demand vanishes), that they increase if excess demand is positive, and decrease if excess demand is negative. The conditions discovered by Arrow and others are that all agents must have perfect foresight for the infinite future (all orders for the future are placed at the initial time, although delivery may occur later as scheduled), and every agent conforms to exactly the same view of the future (the market, which is "complete," is equivalent to the perfect cloning of a single agent as a "utility computer" that can receive all the required economic data, process it, and price all his future demands in a very short time). Here is an example: at time $t = 0$ you plan your entire future, ordering a car on one future date, committing to pay for your kids' education on another date, buying your vacation house on another date, placing all future orders for daily groceries, drugs, long distance charges and gasoline supplies, and heart treatment as well. All demands for your lifetime are planned and ordered in preference. In other words, your and your family's entire future is decided completely at time zero. These assumptions were seen as necessary in order to construct a theory where one could prove rigorous mathematical theorems. Theorem proving about totally unrealistic markets became more important than the empirics of real markets in this picture.

Savings, cash, and financial markets are irrelevant here because no agent needs to set aside cash for an uncertain future. How life should work for real agents with inadequate or uncertain lifelong budget constraints is not and cannot be discussed within the model. In the neo-classical model it is possible to adjust demand schedules somewhat, as new information becomes available, but not to abandon a preplanned schedule entirely.

The predictions of the neo-classical model of an economic agent have proven very appealing to mathematicians, international bankers, and politicians. For example, in the ideal neo-classical world, free of government regulations that hypothetically only promote inefficiency, there is no unemployment. Let L denote the labor supply. With $dL/dt = \varepsilon(L)$, in equilibrium $\varepsilon(L) = 0$ so that everyone who wants to work has a job. This illustrates what is meant by maximum efficiency: no resource goes unused. The introduction of uncertainty in the stationary models of rational expectations avoids 100% efficiency, but provides no insight at all into real macroeconomic problems like inflation and unemployment.

Whether every possible resource (land as community meadow, or public walking path, for example) *ought* to be monetized and used economically is taken for granted, is not questioned in the model, leading to the belief that everything should be priced and traded (see elsewhere the formal idea of Arrow-Debreu prices, a neo-classical notion that foreshadowed in spirit the idea of derivatives). Again, this is a purely postulated abstract theory with no empirical basis, in contrast with real markets made up of qualitatively different kinds of agents with real desires and severe limitations on the availability of information and the ability to sort and correctly interpret information. In the remainder of this chapter we discuss scientific criticism of the neo-classical program from both theoretical and empirical viewpoints, starting with theoretical limitations on optimizing behavior discovered by three outstanding neo-classical theorists.

2.3 The myth of equilibrium via perfect information

In real markets, supply and demand determine nonequilibrium prices. There are bid prices by prospective buyers and ask prices by prospective sellers, so by "price" we mean here the price at which the last trade occurred. This is not a clear definition for a slow-moving, illiquid market like housing, but is well-enough defined for trades of Intel, Dell, or a currency like the Euro, for example. The simplest case for continuous time trading, an idealization of limited validity, would be an equation of the form

$$\frac{dp}{dt} = D(p,t) - S(p,t) = \varepsilon(p,t) \qquad (2.8)$$

where p_k is the price of an item like a computer or a cup of coffee, D is the demand at price p, S is the corresponding supply, and the vector field ε is the excess demand. Phase space is just the n-dimensional p-space, and is flat with no metric (the ps in (2.8) are *always* Cartesian (McCauley, 1997a)). More generally, we could assume that $dp/dt = f(\varepsilon(p,t))$ where f is any vector field with the same qualitative properties as the excess demand. Whatever the choice, we must be satisfied with studying topological classes of excess demand functions, because the excess demand function cannot be uniquely specified by the theory. Given a model, equilibrium is determined by vanishing excess demand, by $\varepsilon = 0$. Stability of equilibrium, when equilibria exist at all, is determined by the behavior of solutions displaced slightly from an equilibrium point. *Note that dynamics requires only that we specify $x = D(p)$, not $p = f(x)$, and likewise for the supply schedule.* The empirical and theoretical importance of this fact will become apparent below.

We must also specify a supply function $x = S(p)$. If we assume that the production time is long on the time scale for trading then we can take the production function to be constant, the "initial endowment," $S(p) \approx x_0$, which is just the total supply at the initial time t_0. This is normally assumed in papers on neo-classical equilibrium theory. In this picture agents simply trade what is available at time $t = 0$, there is no new production (pure barter economy).

With demand assumed slaved to price in the form $x = D(p)$, the phase space is the n-dimensional space of the prices p. That phase space is flat means that global parallelization of flows is possible for integrable systems. The n-component ordinary differential equation (2.8) is then analyzed qualitatively in phase space by standard methods. In general there are $n-1$ time-independent locally conserved quantities, but we can use the budget constraint to show that one of these conservation laws is global: if we form the scalar product of p with excess demand ε then applying the budget constraint to both D and S yields

$$\tilde{p}\varepsilon(p) = 0 \qquad (2.9)$$

The underlying reason for this constraint, called Walras's Law, is that capital and capital accumulation are not allowed in neo-classical theory: neo-classical models assume a pure barter economy, so that the cost of the goods demanded can only equal the cost of the goods offered for sale. This condition means simply that the motion in the n-dimensional price space is confined to the surface of an $n-1$-dimensional sphere. Therefore, the motion is at most $n-1$-dimensional. What the motion looks like on this hypersphere for $n > 3$ is a question that cannot be answered *a priori* without specifying a definite class of models. Hyperspheres in dimensions $n = 3$ and 7 are flat with torsion, which is nonintuitive (Nakahara, 1990). Given a model of excess demand we can start by analyzing the number and character of equilibria and their stability. Beyond that, one can ask whether the motion is integrable. Typically, the motion for $n > 3$ is nonintegrable and may be chaotic or even complex, depending upon the topological class of model considered.

As an example of how easy it is to violate the expectation of stable equilibrium within the confines of optimizing behavior, we present next the details of H. Scarf's model (Scarf, 1960). In that model consider three agents with three assets. The model is defined by assuming individual utilities of the form

$$U_i(x) = \min(x_1, x_2) \qquad (2.10)$$

and an initial endowment for agent number 1

$$x_0 = (1, 0, 0) \qquad (2.11)$$

The utilities and endowments of the other two agents are cyclic permutations on the above. Agent k has one item of asset k to sell and none of the other two assets. Recall that in neo-classical theory the excess demand equation (2.8) is interpreted only as a price-adjustment process, with no trades taking place away from equilibrium. If equilibrium is reached then the trading can only be cyclic with each agent selling his asset and buying one asset from one of the other two agents: either agent 1 sells to agent 2 who sells to agent 3 who sells to agent 1, or else agent 1 sells to agent 3 who sells to agent 2 who sells to agent 1. Nothing else is possible at equilibrium. Remember that if equilibrium is not reached then, in this picture, no trades occur. Also, the budget constraint, which is agent k's income from selling his or her single unit of asset k if the market clears (he or she has no other source of income other than from what he or she sells), is

$$M = \tilde{p}x_0 = p_k \qquad (2.12)$$

Because cyclic trading of a single asset is required, one can anticipate that equilibrium can only be possible if $p_1 = p_2 = p_3$. In order to prove this, we need the idea of "indifference curves."

The idea of indifference curves in utility theory, discussed by I. Fisher (Mirowski, 1989), may have arisen in analogy with either thermodynamics or potential theory. Indifference surfaces are defined in the following way. Let $U(x_1, \ldots x_n) = C$ = constant. If the implicit function theorem is satisfied then we can solve to find one of the xs, say x_i, as a function of the other $n-1$ xs and C. If we hold all xs in the argument of f constant but one, say x_j, then we get an "indifference curve"

$$x_i = f(x_j, C) \qquad (2.13)$$

We can move along this curve without changing the utility U for our "rational preferences." This idea will be applied in an example below.

The indifference curves for agent 1 are as follows. Note first that if $x_2 > x_1$ then $x_1 = C$ whereas if $x_2 < x_1$ then $x_2 = C$. Graphing these results yields as indifference curves $x_2 = f(x_1) = x_1$. Note also that p_3 is constant. Substituting the indifference curves into the budget constraint yields the demand vector components for agent 1 as

$$x_1 = \frac{M}{p_1 + p_2} = D_1(p)$$

$$x_2 = \frac{M}{p_1 + p_2} = D_2(p) \qquad (2.14)$$

$$x_3 = 0$$

2.3 The myth of equilibrium via perfect information

The excess demand for agent 1 is therefore given by

$$\varepsilon_{11} = \frac{p_1}{p_1+p_2} - 1 = -\frac{p_2}{p_1+p_2}$$
$$\varepsilon_{12} = \frac{p_1}{p_1+p_2} \qquad (2.15)$$
$$\varepsilon_{13} = 0$$

where ε_{ij} is the jth component of agent i's excess demand vector. We obtain the excess demands for agents 2 and 3 by cyclic permutation of indices. The kth component of total excess demand for asset k is given by summing over agents

$$\varepsilon_k = \varepsilon_{1k} + \varepsilon_{2k} + \varepsilon_{3k} \qquad (2.16)$$

so that

$$\varepsilon_1 = \frac{-p_2}{p_1+p_2} + \frac{p_3}{p_1+p_3}$$
$$\varepsilon_2 = \frac{-p_3}{p_2+p_3} + \frac{p_1}{p_1+p_2} \qquad (2.17)$$
$$\varepsilon_3 = \frac{-p_1}{p_3+p_1} + \frac{p_2}{p_2+p_3}$$

The excess demand has a symmetry that reminds us of rotations on the sphere. In equilibrium $\varepsilon = 0$ so that

$$p_1 = p_2 = p_3 \qquad (2.18)$$

is the only equilibrium point. It's easy to see that there is a second global conservation law

$$p_1 p_2 p_3 = C_2 \qquad (2.19)$$

following from

$$\varepsilon_1 p_2 p_3 + \varepsilon_2 p_1 p_3 + \varepsilon_3 p_1 p_2 = 0 \qquad (2.20)$$

With two global conservation laws the motion on the 3-sphere is globally integrable, i.e. chaotic motion is impossible (McCauley, 1997a).

It's now easy to see that there are initial data on the 3-sphere from which equilibrium cannot be reached. For example, let

$$(p_{1_0}, p_{2_0}, p_{3_0}) = (1, 1, 1) \qquad (2.21)$$

so that

$$p_1^2 + p_2^2 + p_{31}^2 = 3 \qquad (2.22)$$

Then with $p_{10}p_{20}p_{30} = 1$ equilibrium occurs but for other initial data the plane is not tangent to the sphere at equilibrium and equilibrium cannot be reached. The equilibrium point is an unstable focus enclosed by a stable limit cycle. In general, the market oscillates and cannot reach equilibrium. For four or more assets it is easy to write down models of excess demand for which the motion is chaotic (Saari, 1995).

Suppose that agents have slightly different information initially. Then equilibrium is not computable. That is, the information demands made on agents are so great that they cannot locate equilibrium. In other words, maximum computational complexity enters when we deviate even slightly from the idealized case. It is significant that if agents cannot find an equilibrium point, then they cannot agree on a price that will clear the market. This is one step closer to the truth: real markets are not approximated by the neo-classical equilibrium model. The neo-classical theorist Roy Radner (1968) suggested that liquidity demand, the demand for cash as savings, for example, arises from two basic sources. First, in a certain but still neo-classical world liquidity demand would arise because agents cannot compute equilibrium (although Radner had no idea of a Turing machine, nor did he have a clear idea what he meant by "computational limitations"). Therefore, the agents cannot locate equilibrium. Second, the demand for liquidity should also arise from uncertainty about the future. The notion that liquidity reflects uncertainty appears naturally when we study the dynamics of financial markets; in that case the money bath is the noise created by the traders.

The paper by Bak *et al.* (1999) attempts to define the absolute value of money and is motivated by the fact that a standard neo-classical economy is a pure barter economy, where price p is merely a label[3] as we have described above.

Neo-classical economic theory assumes 100% efficiency (perfect matching a buyer to every seller, and vice-versa), but typical markets outside the financial ones[4] are highly illiquid and inefficient (housing, automobiles, floor-lamps, carpets, etc.) where it is typically relatively hard to match buyers to sellers. Were it easy to match buyers to sellers, then advertising and inventory would be largely superfluous. Seen from this standpoint, one might conclude that advertising may distort markets instead of making them more efficient. Again, it would be important to distinguish advertising as formal "information" from knowledge of empirical facts. In financial markets, which are

---

[3] In a standard neo-classical economy there is no capital accumulation, no financial market, and no production of goods either. There is merely exchange of pre-existing goods.
[4] Financial markets are far from 100% efficient; excess demand does not vanish due to outstanding limit orders.

2.3 The myth of equilibrium via perfect information

usually very liquid (with a large volume of buy and sell executions per second), the neo-classical economic assumptions of equilibrium and stability fail even as a zeroth-order approximation.

The absence of entropy representing disorder in neo-classical equilibrium theory can be contrasted with thermodynamics in the following way: for assets in a market let us define economic efficiency as

$$e = \min\left(\frac{D}{S}, \frac{S}{D}\right) \qquad (2.23)$$

where S and D are net supply and net demand for some asset in that market. In neo-classical equilibrium the efficiency is 100%, $e = 1$, whereas the second law of thermodynamics via the heat bath prevents 100% efficiency in any thermodynamic machine. That is, the neo-classical market equilibrium condition $e = 1$ is not a thermodynamic efficiency, unless we would be able to interpret it as the zero (Kelvin) temperature result of an unknown thermodynamic theory (100% efficiency of a machine is thermodynamically possible only at zero absolute temperature). In nature or in the laboratory, superfluids flow with negligible friction below the lambda temperature, and with zero friction at zero degrees Kelvin, at speeds below the critical velocity for creating a vortex ring or vortex pair. In stark contrast, neo-classical economists assume the unphysical equivalent of a hypothetical economy made up of Maxwellian demonish-like agents who can systematically cheat the second law perfectly. We should add that the attempts at thermodynamic descriptions of economics are many, and all are wrong. There is no entropy/disorder in neo-classical economics.

In neo-classical equilibrium theory perfect information about the infinite future is required and assumed. In reality, information acquired at one time is incomplete and tends to become degraded as time goes on. Entropy change plays no role in neo-classical economic theory in spite of the fact that, given a probability distribution reflecting the uncertainty of events in a system (the market), the Gibbs entropy describes both the accumulation and degradation of information. Neo-classical theory makes extreme demands on the ability of agents to gather and process information, but as Fischer Black wrote, it is extremely difficult in practice to know what is noise and what is information (we will discuss Black's 1986 paper "Noise" in Chapter 4). For example, when one reads the financial news one usually only reads someone else's opinion, or assertions based on assumptions that the future will be more or less like the past. Most of the time, what we think is information is probably more like noise or misinformation. This point of view is closer to finance theory, which does not use neo-classical economics as a starting point.

Another important point is that information should not be confused with knowledge (Dosi, 2001). The symbol string "saht" (based on at least a 26-letter alphabet a–z) has four digits of information, but without a rule to interpret it the string has no meaning, no knowledge content. In English we can give meaning to the combinations "hast," "hats," and "shat." Information theory is based on the entropy of all possible strings that one can make from a given number of symbols, that number being 4! = 24 in this example, but "information" in standard economics and finance theory does not make use of entropy.

Why would neo-classical thinking be useful were it correct? Because the equilibrium label p* (a star denotes equilibrium) could be identified as "value." Undervalued and overvalued would then be well-defined, measurable ideas. Unfortunately, p* could not be identified as an equilibrium price because money does not enter the theory. We'll see in Chapter 4 that a more useful definition of value can be identified in uncertain markets.

2.4 How many green jackets does a consumer want?

An empirically based criticism of neo-classical theory was provided by M. F. M. Osborne (1997), whom we can regard as the first econophysicist. According to the standard textbook argument, utility maximization for the case of diminishing returns predicts price as a function of demand, $p = f(x)$, as a downward-sloping curve (Figure 2.2). Is there empirical evidence for this prediction? Osborne tried without success to find empirical evidence for the textbook supply and demand curves (Figure 2.3), whose intersection would determine equilibrium. This was an implicit challenge to the notion that markets are in or near equilibrium. In the spirit of Osborne's toy model of a market for red dresses, we now provide a Gedanken experiment to illustrate how the neo-classical prediction fails for individual agents. Suppose that I'm in the market for a green jacket. My neo-classical demand curve would then predict that I, as consumer, would have the following qualitative behavior, for example: I would want/bid to buy one green jacket for $50, two for $42.50 each, three for $31.99 each, and so on (and this hypothetical demand curve would be continuous!). Clearly, no consumer thinks this way. This is a way of illustrating Osborne's point, that the curve $p = f(x)$ does not exist empirically for individual agents.

What exist instead, Osborne argues, are the functions $x = D(p)$ and $x = S(p)$, which are exactly the functions required for excess demand dynamics (2.8). Osborne notes that these functions are not invertible, implying that utility cannot explain real markets. One can understand the lack of invertibility by modeling my demand for a green jacket correctly. Suppose that I want one

2.5 Macroeconomics

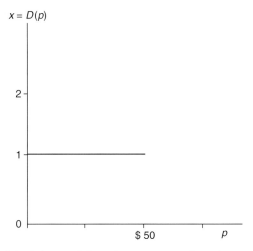

Figure 2.4 Empirical demand functions are step functions.

jacket and am willing to pay a maximum of $50. In that case I will take any (suitable) green jacket for $50 or less, so that my demand function is a step function $x = \theta\, (\$50 - p)$, as shown in Figure 2.4. The step function θ is zero if $p > \$50$, unity if $p \leq \$50$. Rarely, if ever, is a consumer in the market for *two* green jackets, and one is almost never interested in buying three or more at one time. Nevertheless, the step function can be used to include these rare cases. This argument is quite general: Osborne points out that limit bid/ask orders in the stock market are also step functions (one can see this graphically in delayed time on the web site 3DStockCharts.com). Limit orders and the step demand function for green jackets provide examples of the falsification of the neo-classical prediction that individual agents have downward-sloping demand curves $p = f(x)$. With or without equilibrium, the utility-based prediction is wrong. Optimizing behavior does not describe even to zeroth order how individual agents order their preferences. Alternatives like wanting one or two of several qualitatively different jackets can also be described by step functions just as limit orders for different stocks are described by different step functions. The limit order that is executed first wins, and the other orders are then cancelled unless there is enough cash for more than one order.

2.5 Macroeconomics

One might raise the following question: suppose that we take many step functions $x = D(p)$ for many agents and combine them. Do we get approximately a smooth curve that we can invert to find a relation $p = f(x)$ that agrees qualitatively with the downward-sloping neo-classical prediction? In agreement with Osborne's attempts, apparently not empirically, the economist Paul Ormerod has pointed out that the only known downward-sloping macroscopic demand curve is provided by the example of cornflakes sales in British supermarkets.

What about theory? If we assume neo-classical individual demand functions and then aggregate them, do we arrive at a downward-sloping macro-demand curve? According to H. Sonnenschein (1973) the answer is no, that no definite demand curve is predicted by aggregation; the resulting curve can be anything, including no curve at all. In other words, nothing definite is predicted. This means that there exists no macroeconomic theory that is grounded in microeconomic theory. What is worse, there is no empirical evidence for the downward-sloping demand curves presented in typical neo-classical texts on macroeconomics, like the relatively readable one by N. G. Mankiw (2000). This means that there is no microeconomic basis for either Keynesian economics or monetarism, both of which make empirically illegitimate assumptions about equilibrium.

For example, in Keynesian theory (Modigliani, 2001) it is taught that there is an aggregate output equilibrium where the labor market is "stuck" at less than full employment but prices do not drop as a consequence. Keynes tried to explain this via an equilibrium model that went beyond the bounds of neo-classical reasoning. The neo-classicals led by J. R. Hicks later revised theoretical thinking to try to include neo-Keynesianism in the assumption of vanishing total excess demand for all goods and money, but Radner has explained why money cannot be included meaningfully in the neo-classical model. A better way to understand Keynes's original idea is to assume that the market is not in equilibrium,

$$\frac{dp}{dt} = \varepsilon_1(p, L) \neq 0 \tag{2.22b}$$

with

$$\frac{dL}{dt} = \varepsilon_2(p, L) \neq 0 \tag{2.22c}$$

where p is the price vector of commodities and financial markets. But a deterministic model will not work: financial markets (which are typically highly liquid) are described by stochastic dynamics. Of interest would be to model the Keynesian liquidity trap (see Ackerlof, 1984; Krugman, 2000) without assuming expected utility maximization. There, one models markets where liquidity dries up. If one wants to model nonequilibrium states that persist for long times, then maybe spin glass/neural network models would be interesting.

John Maynard Keynes advanced a far more realistic picture of markets than do monetarists by arguing that capitalism is not a stable, self-regulating system capable of perpetual prosperity. Instead, he saw markets as inherently unstable, occasionally in need of a fix by the government. We emphasize that by neglecting uncertainty the neo-classical equilibrium model ignores the second law prohibition against the construction of an economic perpetuum mobile. The idea of a market as frictionless, 100% efficient machine (utility computer) that runs perpetually is a wrong idea from the standpoint of statistical physics. Markets require mechanical acts like production, consumption, and information gathering and processing, and certainly cannot evade or supplant the second law of thermodynamics simply by postulating utility maximization. Keynes's difficulty in explaining his new and important idea was that while he recognized the need for the idea of nonequilibrium markets in reality, his neo-classical education mired him in the sticky mud of equilibrium ideas. Also, his neo-classical contemporaries seemed unable to understand any economic explanation that could not be cast into the straitjacket of an equilibrium description.

Monetarism (including supply-side economics) and Keynesian theory are examples of ideas that have become ideologies, because they both represent attempts to use equilibrium arguments to describe the behavior and regulation of a complex system by controlling the parameters of inapplicable models. The advice provided by both approximations was found to be useful by governments during certain specific eras (otherwise they would not have become widely believed), but all of that advice has failed in our present era of high inflation and loss of manufacturing capacity due to globalization via deregulation. In monetarism one controls the money supply, in Keynesianism the level of government spending, while in the supply-side belief tax reductions dominate the thinking. We will consider Keynesian and monetarist notions further in Chapter 10.

Marxism and other earlier competing economic theories of the nineteenth and early twentieth centuries also assumed stable equilibria of various kinds. In Marxism, for example, evolution toward a certain future is historically guaranteed. This assumption is equivalent mathematically to assuming a stable fixed point, a simple attractor for some undetermined mapping of society. Society was supposed somehow to iterate itself toward this inevitable state of equilibrium with no possible choice of any other behavior, a silly assumption based on wishful thinking. But one of Karl Marx's positive contributions was to remind us that the neo-classical model ignores the profit motive completely: in a pure barter economy the accumulation of capital is impossible, but capitalists are driven to some extent by the desire to accumulate capital. Marx reconnected economic theory to Adam Smith's original idea of the profit motive.

Evidence for stability and equilibrium in unregulated markets is largely if not entirely anecdotal, more akin to excessively weak circumstantial evidence in legal circles than to scientific evidence. Convincing, reproducible empirical evidence for the Invisible Hand has never been presented by economists. Markets whose statistics are well-enough defined to admit description by falsifiable stochastic models (financial markets) are unstable (see Chapter 7). It would be an interesting challenge to find at least one example of a real, economically significant market where excess demand actually vanishes and remains zero or close to zero to within observational error, where only small fluctuations occur about a definite state of equilibrium. A flea market is an example where equilibrium is never reached. Some trades are executed but at the end of the day most of the items put up for sale are carried home again because most ask prices were not met, or there was inadequate demand for most items. Selling a few watches from a table covered with watches is not an example of equilibrium or near-equilibrium. The same goes for filling a fraction of the outstanding limit orders in the stock market.

We now summarize the evidence from the above sections against the notion that equilibrium exists, as is assumed explicitly by the intersecting neo-classical supply–demand curves shown in Figure 2.3. Scarf's model shows how easy it is to violate stability of equilibrium with a simple model. Sonnenschein explained that neo-classical supply–demand curves cannot be expected macroeconomically, even if they would exist microeconomically. Osborne explained very clearly why neo-classical supply–demand curves do not exist microeconomically in real markets. Radner showed that with even slight uncertainty, hypothetical optimizing agents cannot locate the equilibrium point assumed in Figure 2.3, even in a nearly ideal, toy neo-classical economy. And yet, intersecting neo-classical supply–demand curves remain the foundation of nearly every standard economics textbook. See also Keen (2001) and Ormerod (1994) for discussions of the neo-classical model.

Finally, the notion that free trade among nations is self-stabilizing is a neo-classical idea, taking nations as agents. We will return to this theme in Chapters 9 and 10, where the more modern notion of rational expectations generalizes the neo-classical model to include noise.

3

Probability and stochastic processes

3.1 Elementary rules of probability theory

The aim of this chapter is to prepare the reader to understand the stochastic analysis used to model finance and other macroeconomic data, and to prepare the reader for the analysis of nonstationary time series via statistical ensemble analysis in Chapter 7. This chapter provides the mathematics needed to follow the rest of the text.

In the absence of the laws of physics, which were themselves extracted from nature, the extraction from empirical data is the only scientific basis for a model. We therefore adopt from the start the frequency definition of probability based upon the law of large numbers, Tschebychev's Theorem, which is presented below. The frequency definition of probability is also called the empirical definition, or the statistical definition. Given an event with possible outcomes $A_1, A_2, \ldots, A_N$, the probability for A_k is $p_k \approx n_k/N$ where N is the number of repeated identical experiments or observations and n_k is the number of times that the event A_k is observed to occur. The statistical definition of probability agrees with the formal measure theoretic definition. For equally probable events $p = 1/N$.

For mutually exclusive events (Gnedenko and Khinchin, 1962; Gnedenko, 1967) A and B, probabilities add, $P(A \text{ or } B) = P(A) + P(B)$. For example, the probability that a coin lands heads plus the probability that it does not land heads add to unity (total probability is normalized to unity in this text). For a complete (i.e. exhaustive) set of mutually exclusive alternatives $\{A_k\}$, we have $\Sigma P(A_k) = 1$. For example, in die tossing, if p_k is the probability for the number k to show, where $1 \leq k \leq 6$, then $p_1 + p_2 + p_3 + p_4 + p_5 + p_6 = 1$. For a fair die tossed fairly, $p_k = 1/6$.

For statistically independent events A and B, the probabilities multiply, $P(A \text{ and } B) = P(A)P(B)$, *and this is true for all combinations of multiple events as well*. For example, for n successive fair tosses of a fair coin ($p = 1/2$) the probability to get n heads is $p^n = (1/2)^n$. Statistical independence is often mislabeled as "randomness," but statistical independence occurs in deterministic chaos where there is no randomness at all; there is in that case only the pseudorandom generation of numbers deterministically via an algorithm.

We can use what we have developed so far to calculate a simple formula for the occurrence of at least one desired outcome in many events. For this, we need the probability that the event does not occur. Suppose that p is the probability that event A occurs. Then the probability that the event A does not occur is $q = 1 - p$. The probability to get at least one occurrence of A in n repeated identical trials is $1 - q^n$. As an example, the probability to get at least one "six" in n tosses of a fair (where $p = 1/6$) die is $1 - (5/6)^n$. The break-even point is given by $1/2 = (5/6)^n$, or $n \approx 4$ is required to break even. One can make money by getting many people to bet that a 6 won't occur in 4 (or more) tosses of a die so long as one does not suffer the Gambler's Ruin (so long as an unlikely run against the odds doesn't break your gambling budget). That is, we should not only consider the expected outcome of an event or process, we must also look at the fluctuations.

What are the odds that at least two people in one room have the same birthday? We leave it to the reader to show that the break-even point (even odds) for the birthday game requires $n = 22$ people (Weaver, 1982). The method of calculation is the same as in the paragraph above.

Stock market betting is not as simple as standard examples with probabilities known *a priori*. As Keynes stated, finance market success is more like a beauty contest where in order to win one must anticipate which candidate the other players will find most beautiful before the voting takes place. The popularity contest over a short time interval can determine the financing that affects the chance whether a company fails or succeeds in the long term as well.

3.2 Ensemble averages formed empirically

Consider any collection of n points arranged on the x-axis, $x_1, x_2, \ldots, x_n$. The "points" are known empirically to within some definite decimal precision, and that decimal precision determines the coarsegraining of the x-axis into cells (also called boxes, or bins). For example, if the observational precision is known to within .01, then there are a hundred boxes in [0,1], each of size .01, and the normalized data would fall into one or another of those boxes. Naturally, some boxes may be multiply occupied. We can define a distribution for the point set. Let $P(x)$ denote the probability that a point

3.2 Ensemble averages formed empirically

lies to the left of x on the x-axis. The empirical (one-point) probability distribution is then

$$P(x) = \sum_{i=1}^{k} \theta(x - x_i)/n \tag{3.1}$$

where x_k is the nearest point to the left of x, $x_k \leq x$ and $\theta(x) = 1$ if $x \geq 0$, 0 otherwise. Note that $P(-\infty) = 0$ and $P(\infty) = 1$. The function $P(x)$ is nondecreasing, defines a staircase of a finite number of steps, is constant between any two data points and is discontinuous at each data point. $P(x)$ satisfies all of the formal conditions required to define a probability measure mathematically. Theoretical measures like the Cantor function define a probability distribution on a staircase of infinitely many steps, a so-called devil's staircase.

There is no variation with time here (no dynamics), such a point set arises empirically by defining an ensemble of data points at one fixed time t. The ensemble represents many identical repetitions of the same experiment under the same conditions. That is, for a given time t if we would only consider one run of the experiment then we would only have one point in (3.1). That is, one run of an experiment produces one point in the time series $x_1(t)$ at time t. n reruns, under identical experimental conditions, produce n different time series $x_k(t)$, $k = 1,\ldots,n$ at the same time t (we assume, for example, the use of a 24-hour clock with each experiment repeated at the same time each day). The averages that we discuss represent an ensemble average over n points taken *at the same time t*. The histograms so derived represent the probability distribution at a single time t. To discover the time variation of averages and the distribution, the procedure must be repeated for every time t.

In the first few sections to follow, where the time dependence is not made explicit, averages and distributions are assumed to reflect ensemble averages computed at one time t. We will not assume time averages of any kind in this text, especially as we're concerned with nonstationary processes where ergodic and recurrence theorems fail to apply.

We can also construct the probability density $f(x)$ where $dP(x) = f(x)dx$

$$f(x) = \sum_{i=1}^{n} \delta(x - x_i)/n \tag{3.2}$$

We can then compute averages using the empirical distribution. For example,

$$\langle x \rangle = \int_{-\infty}^{\infty} x dP(x) = \frac{1}{n}\sum_{1}^{n} x_i \tag{3.3}$$

and

$$\langle x^2 \rangle = \int_{-\infty}^{\infty} x^2 \mathrm{d}P(x) = \frac{1}{n}\sum_1^n x_i^2 \qquad (3.4)$$

The variance is defined by

$$\sigma^2 = \langle \Delta x^2 \rangle = \left\langle (x - \langle x \rangle)^2 \right\rangle = \langle x^2 \rangle - \langle x \rangle^2 \qquad (3.5)$$

The standard deviation σ is taken as an indication of the usefulness of the average (3.3) for characterizing the data. The data are accurately characterized by the mean if and only if

$$\sigma/|\langle x \rangle| \ll 1 \qquad (3.6)$$

and even then only for a sequence of many identical repeated experiments or approximately identical repeated observations.

Statistics generally have no useful predictive power for a single experiment or observation, and can at best be relied on for accuracy in predictive power for an accurate description of the average of many repeated trials. In social applications where nothing can be repeated at will, like the observed price of a stock, we have but a single historic time series. The problem addressed in Chapter 7 is the question of whether and how an approximate statistical ensemble can be constructed from a single time series.

3.3 The characteristic function

An idea that's sometimes useful in probability theory is that of the characteristic function of a distribution, defined by the Fourier transform

$$\langle \mathrm{e}^{ikx} \rangle = \int \mathrm{d}P(x) \mathrm{e}^{ikx} \qquad (3.7)$$

Expanding the exponential in power series we obtain the expansion in terms of the moments of the distribution

$$\langle \mathrm{e}^{ikx} \rangle = \sum_{m=0}^{\infty} \frac{(ik)^m}{m!} \langle x^m \rangle \qquad (3.8)$$

showing that the distribution is characterized by all of its moments (with some exceptions), and not just by the average and variance. For an empirical distribution the characteristic function has the form

3.4 Transformations of random variables

$$\langle e^{ikx} \rangle = \sum_{j=1}^{n} e^{ijkx_j}/n \qquad (3.9)$$

Clearly, if all moments beyond a certain order m diverge (as with Levy distributions, for example) then the expansion (3.9) of the characteristic function does not exist.

Empirically, smooth distributions do not exist. Only histograms can be constructed from data, but we will still consider model distributions $P(x)$ that are smooth with continuous derivatives of many orders, $dP(x) = f(x)dx$, so that the density $f(x)$ is at least once differentiable. Smooth distributions are useful if they can be used to approximate observed histograms accurately.

3.4 Transformations of random variables

In order to perform simple coordinate transformations, one must first know how probabilities and probability densities transform. Here, we use standard ideas from tensor analysis or group theory.

In the smooth case, transformations of the variable x are important. Consider a transformation of a variable $y = h(x)$ with inverse $x = q(y)$. The new distribution of y has density

$$\tilde{f}(y) = f(x)\frac{dy}{dx} \qquad (3.10)$$

For example, if

$$f(x) = e^{-x^2/2\sigma^2} \qquad (3.11)$$

where $x = \ln(p/p_0)$ and $y = (p - p_0)/p_0$ then $y = h(x) = e^x - 1$ so that

$$\tilde{f}(y) = \frac{1}{1+y} e^{-(\ln(1+y))^2/2\sigma^2} \qquad (3.12)$$

The probability density transforms $f(x)$ like a *scalar density*, and the probability distribution $P(x)$ transforms like a scalar (i.e. like an ordinary function),

$$\tilde{P}(y) = P(x) \qquad (3.13)$$

Whenever a distribution is *invariant* under the transformation $y = h(x)$ then

$$P(y) = P(x) \qquad (3.14)$$

That is, the functional form of the distribution doesn't change under the transformation. As an example, if we replace p and p_0 by λp and λp_0, a scale transformation, then neither an arbitrary density $f(x)$ nor its corresponding distribution *P(x)* is invariant. In general, even if *f(x)* is invariant then *P(x)* is not, unless both d*x* and the limits of integration in

$$P(x) = \int_{-\infty}^{x} f(x)dx \qquad (3.15)$$

are invariant. The distinction between scalars, scalar densities, and invariants is stressed here, because even books on relativity often write "invariant" when they should have written "scalar" (Hamermesh, 1962; McCauley, 2001).

3.5 Laws of large numbers

We will need to distinguish in all that follows between statistically independent variables and merely pairwise uncorrelated variables. Often, writers assume statistical independence whereas in fact all that is used is lack of pairwise correlation. In principle and in practice, the difference is enormous, as certain combinations of variables in data analysis and modeling are often uncorrelated but are seldom or never statistically independent.

A set of *n* random variables is statistically independent if the joint *n*-point density factors into *n* one-point densities,

$$f_n(x_n, \ldots, x_1) = f_n(x_n) \ldots f_1(x_1). \qquad (3.16)$$

If the *n* one-point densities are the same, if $f_k(x_k) = f_1(x_k)$ for $k = 2, \ldots, n$, then the *n* variables are called independently identically distributed, or "iid." A far weaker but much more useful condition is that the *n* variables are not statistically independent but are pairwise uncorrelated, i.e. $\langle x_j x_k \rangle = 0$ if $j \neq k$. An example is provided by the increments (displacements) in a discrete or continuous random walk, where the positions themselves are correlated.

3.5.1 Tschebychev's inequality

Tschebychev's inequality states that for every random variable *x* with finite mean $\langle x \rangle = a$ and variance σ^2, the inequality

$$P(|x-a| \geq \varepsilon) \leq \sigma^2/\varepsilon^2 \qquad (3.17)$$

holds for every $\varepsilon > 0$. The proof is easy:

$$P(|x-a| \geq \varepsilon) = \int_{|x-a|\geq\varepsilon} dP(x) \leq \frac{1}{\varepsilon^2} \int_{|x-a|\geq\varepsilon} (x-a)^2 dP(x) = \frac{\sigma^2}{\varepsilon^2} \qquad (3.18)$$

3.5 Laws of large numbers

We assume here that, even if the distribution has fat tails, the tails are not fat enough to make the variance blow up. An example is presented in a later section. We can choose $\varepsilon = n\sigma$, obtaining as the probability for an "$n\sigma$-event"

$$P(|x - a| \geq n\sigma) \leq \frac{1}{n^2} \tag{3.19}$$

This result has interesting practical implications. Irrespective of the distribution, the probability of a 2σ event is bounded only by ¼, which is a large probability. Were the distribution Gaussian we could achieve a much tighter bound, but finance distributions are not Gaussian. No assumption of stationarity has been made here. The result applies to the nonstationary distributions of finance where σ increases with time.

Next, we introduce a result of much practical use. Tschebychev's Theorem provides the justification for the construction of statistical ensembles using repeated identical experiments. Here, statistics and laboratory science merge in a very fruitful marriage. Although the theorem requires variables with a common mean, which would be hard to satisfy, we will apply the result to detrended variables in finance in Chapter 7.

3.5.2 Tschebychev's Theorem

Assume a large number of pairwise uncorrelated variables $x_1, \ldots, x_n$ with the same ensemble average mean a, and uniformly bounded variances.

Let x_k occur p_k times with $k = 1, \ldots, m$. Then

$$\langle x \rangle = \int x dP(x) = \frac{1}{n} \sum_{j=1}^{n} x_j = \sum_{k=1}^{m} p_k x_k \tag{3.20}$$

Define a new random variable x as

$$x = \frac{1}{n} \sum_{k=1}^{n} x_k \tag{3.21}$$

From

$$x - \langle x \rangle = \frac{1}{n} \sum_{j=1}^{n} (x_j - \langle x \rangle) \tag{3.22}$$

we obtain

$$(x - \langle x \rangle)^2 = \frac{1}{n^2} \sum_{j=1}^{n} (x_j - \langle x \rangle)^2 + \frac{1}{n^2} \sum_{j \neq k}^{n} (x_j - \langle x \rangle)(x_k - \langle x \rangle) \tag{3.23}$$

so that

$$\sigma_x^2 = \frac{1}{n^2}\sum_{j=1}^{n}\left\langle (x_j - \langle x\rangle)^2\right\rangle = \frac{1}{n^2}\sum_{j=1}^{n}\sigma_j^2 \leq \frac{(\sigma_j^2)_{\max}}{n} \quad (3.24)$$

where

$$\sigma_j^2 = \left\langle (x_j - \langle x_j\rangle)^2\right\rangle \quad (3.25)$$

Tschebychev's inequality then yields

$$P(|x - \langle x\rangle| > \alpha) \leq \frac{(\sigma_j^2)_{\max}}{n\alpha^2} \quad (3.26)$$

This is the main result of general interest for us later.

If, in addition, the n random variables would have the same variance σ^2 then we obtain from (3.24) that

$$\sigma_x^2 = \frac{\sigma^2}{n} \quad (3.27)$$

which also suggests that scatter can be reduced by studying the sum x of n uncorrelated variables instead of the individual variables x_k.

The law of large numbers requires only pairwise uncorrelated variables. The central limit theorem (CLT) described next is less applicable, as it requires the stronger and empirically unrealistic assumption of statistical independence. The CLT can be used to prove powerful convergence theorems in probability that do not apply widely enough to empirical data. The law of large numbers, however, is quite general and even provides the basis for constructing statistical ensemble averages based on repeated identical experiments.

3.5.3 The central limit theorem

We showed earlier that a probability distribution $P(x)$ may be characterized by its moments via the characteristic function $F(k)$, which we introduced in part 3.3 above. The Fourier transform of a Gaussian is again a Gaussian,

$$\phi(k) = \frac{1}{\sqrt{2\pi}\sigma}\int_{-\infty}^{\infty} dx\, e^{ikx} e^{-(x-\langle x\rangle)^2/2\sigma^2} = e^{ik\langle x\rangle} e^{-k^2\sigma^2/2} \quad (3.28)$$

We now show that the Gaussian plays a special role in a certain ideal limit. Consider N independent random variables x_k, which may or may not be identically distributed. Each has finite standard deviation σ_k. That is, the

3.5 Laws of large numbers

individual distributions $P_k(x_k)$ need not be identical; the central assumption is statistical independence. We can formulate the problem in either of two ways.

We may ask directly what is the distribution $P(x)$ of the variable

$$x = \frac{1}{\sqrt{n}} \sum_{k=1}^{n} x_k \tag{3.29}$$

where we can assume that each x_k has been constructed to have vanishing mean. The characteristic function is

$$\Phi(k) = \int_{-\infty}^{\infty} e^{ikx} dP(x) = \langle e^{ikx} \rangle = \left\langle \prod_{k=1}^{n} e^{ikx_k/\sqrt{n}} \right\rangle = \prod_{k=1}^{n} \left\langle e^{ikx_k/\sqrt{n}} \right\rangle \tag{3.30}$$

where statistical independence was used in the last step. Writing

$$\Phi(k) = \langle e^{ikx} \rangle = \prod_{k=1}^{n} \left\langle e^{ikx_k/\sqrt{n}} \right\rangle = e^{\sum_{k=1}^{n} A_k(k/\sqrt{n})} \tag{3.31}$$

where

$$A_k(k/\sqrt{n}) = \ln \left\langle e^{ikx_k/\sqrt{n}} \right\rangle \tag{3.32}$$

we can expand to obtain

$$A_k(k/\sqrt{n}) = A_k(0) + k^2 A_k''(0)/2n + k^3 O(n^{-1/2})/n + \ldots \tag{3.33}$$

where

$$A_k''(0) = \langle x_k^2 \rangle \tag{3.34}$$

If, as N goes to infinity, we could neglect terms of order k^3 and higher in the exponent of $F(k)$ then we would obtain the Gaussian limit

$$\langle e^{ikx} \rangle = e^{\sum A_k(k/\sqrt{N})} \approx e^{-k^2 \sigma_x^2/2} \tag{3.35}$$

where σ_x is the variance of the cumulative variable x.

An equivalent way to derive the same result is to start with the convolution of the individual distributions subject to the constraint (3.29)

$$P(x) = \int \ldots \int dP_1(x_1) \ldots dP_n(x_n) \delta(x - \sum x_k/\sqrt{n}) \tag{3.36}$$

Using the Fourier transform representation of the delta function yields

$$\Phi(k) = \prod_{i=1}^{N} \phi_i(k/\sqrt{N}) \tag{3.37}$$

where f_k is the characteristic function of P_k, and provides another way to derive the CLT.

A nice example that shows that limitations of the CLT is provided by Bouchaud and Potters (2000) who consider the asymmetric exponential density

$$f_1(x) = \theta(x)\alpha e^{-\alpha x} \qquad (3.38)$$

Using (3.40) in (3.36) yields the density

$$f(x,N) = \theta(x)\alpha^N \frac{x^{N-1}e^{-\alpha x}}{(N-1)!} \qquad (3.39)$$

Clearly, this distribution is never Gaussian for either arbitrary or large values of x. What, then, does the CLT describe? If we locate the value of x for which $f(x,N)$ is largest, the most probable value of x, and approximate $\ln f(x,N)$ by a Taylor expansion to second order about that point, then we obtain a Gaussian approximation to f. Since the most probable and mean values approximate each other for large N, we see that the CLT asymptotically describes *small* fluctuations about the mean. However, the CLT does not describe the distribution of very small or very large values of x correctly for *any* value of N. Even worse, the common expectation that a financial returns distribution should become Gaussian at long times is completely misplaced: we'll see that neither financial returns nor returns differences satisfy the conditions for statistical independence.

In this text we will not appeal to the CLT in data analysis because that theorem does not provide a reasonable approximation for a large range of values of x. It is possible to go beyond the CLT and develop formulae for "large deviations" and "extreme values," but we will not need those results in this text and so refer the interested reader to the literature (Frisch and Sornette, 1997).

3.6 Examples of theoretical distributions

The Gaussian and lognormal distributions (related by a coordinate transformation) form the basis for financial engineering. The exponential distribution has been discovered in finance, hard turbulence, and firm growth data. Both student-*t*-like and Levy densities exhibit fat tails, but have entirely different origins dynamically. Student-*t*-like densities are derived in Chapter 6 from diffusive processes. Levy densities do not satisfy a diffusion equation because the variance is infinite.

3.6.1 Gaussian and lognormal densities

A Gaussian distribution has the one-point density

$$f(x) = \frac{1}{\sqrt{2\pi}\sigma} e^{-(x-\langle x \rangle)^2/2\sigma^2} \qquad (3.40)$$

with variance

$$\sigma^2 = \langle \Delta x^2 \rangle \qquad (3.41)$$

and plays a special role in probability theory. On the one hand it arises as a limit distribution from the law of large numbers. On the other hand, a special Gaussian process, the Wiener process, forms the basis for stochastic calculus. If we take $x = \ln p$ then $g(p)dp = f(x)dx$ defines the density $g(p)$, which is lognormal in the variable p. The lognormal distribution was first applied in finance by Osborne in 1958 (Cootner, 1964), and was used later by Black, Scholes, and Merton in 1973 to price options falsifiably via a special trading strategy.

3.6.2 The exponential density

The asymmetric exponential density (Laplace density) was discovered in an analysis of financial data by Gunaratne c. 1990. The exponential density survived in finance analysis even after a more careful data analysis was performed (Chapter 7).

A version of the asymmetric exponential density is defined by

$$f(x) = \begin{cases} \frac{\gamma}{2} e^{\gamma(x-\delta)} & x < \delta \\ \frac{\nu}{2} e^{-\nu(x-\delta)} & x > \delta \end{cases} \qquad (3.42)$$

where δ, γ, and ν are the parameters that define the distribution and generally depend on time. Many different possible normalizations of the density are possible. The normalization chosen above is not the one required to conserve probability in a stochastic dynamical description. That normalization is derived in Chapter 8.

Moments of this distribution are easy to calculate in closed form. For example,

$$\langle x \rangle_+ = \int_\delta^\infty x f(x) dx = \delta + \frac{1}{\nu} \qquad (3.43)$$

is the mean of the distribution for $x > \delta$ while

$$\langle x \rangle_- = \int_{-\infty}^{\delta} xf(x)dx = \delta - \frac{1}{\gamma} \tag{3.44}$$

defines the mean for that part with $x < \delta$. The mean of the distribution is given by

$$\langle x \rangle = \delta + \frac{(\gamma - \nu)}{\gamma \nu} \tag{3.45}$$

The analogous expressions for the mean square are

$$\langle x^2 \rangle_+ = \frac{2}{\nu^2} + 2\frac{\delta}{\nu} + \delta^2 \tag{3.46}$$

and

$$\langle x^2 \rangle_- = \frac{2}{\gamma^2} - 2\frac{\delta}{\gamma} + \delta^2 \tag{3.47}$$

Hence the variances for the distinct regions are given by

$$\sigma_+^2 = \frac{1}{\nu^2} \tag{3.48}$$

$$\sigma_-^2 = \frac{1}{\gamma^2} \tag{3.49}$$

and for the whole by

$$\sigma^2 = \frac{\gamma^2 + \nu^2}{\gamma^2 \nu^2} \tag{3.50}$$

We can estimate the probability of large events. The probability for at least one event $x > \sigma$ is given (for $x > \delta$) by

$$P(x > \sigma) = \frac{\nu}{2} \int_{\sigma}^{\infty} e^{-\nu(x-\delta)} dx = \frac{1}{2} e^{-\nu(\sigma-\delta)} \tag{3.51}$$

The exponential density was observed in foreign exchange (FX) data (and also in hard turbulence) by Gunaratne (McCauley and Gunaratne, 2003) and in the analysis of firm size growth rates by Bottazzi and Secchi (2005) and by Lee et al. (1998).

3.6.3 Student-t-like densities (fat tails)

Pareto introduced fat-tailed densities in wealth accumulation studies in the nineteenth century. Fat-tailed densities do not arise from a single, unique dynamic model. Instead, they arise from various mutually exclusive models. One class is student-*t*-like.

An interesting class of student-*t*-like densities is given by

$$f(x) = [1 + \varepsilon x^2]^{-1-\alpha}. \tag{3.52}$$

This density has fat tails,

$$f(x) \approx |x|^{-\mu}, |x| \gg 1 \tag{3.53}$$

where $\mu = 2 + 2\alpha$ is the tail exponent. Consequently all moments $\langle x^n \rangle$ blow up for $n > 1 + 2\alpha$. The variance $\sigma^2 = \langle x^2 \rangle$ is finite if $\alpha > 1/2$. Student-*t*-like densities can be generated by diffusive dynamics.

3.6.4 Stretched exponential distributions

The stretched exponential density is defined by

$$f(x,t) = \begin{cases} Ae^{-(\nu(x-\delta))^\alpha}, & x > \delta \\ Ae^{(\gamma(x-\delta))^\alpha}, & x < \delta \end{cases} \tag{3.54}$$

Using

$$dx = \nu^{-1} z^{1/\alpha - 1} dz \tag{3.55}$$

we can easily evaluate all averages of the form

$$\langle z^n \rangle_+ = A \int_\delta^\infty (\nu(x-\delta))^{n\alpha} e^{-(\nu(x-\delta))^\alpha} dx \tag{3.56}$$

for n an integer. Therefore we can reproduce analogs of the calculations for the exponential distribution. For example,

$$A = \frac{\gamma \nu}{\gamma + \nu} \frac{1}{\Gamma(1/\alpha)} \tag{3.57}$$

where $\Gamma(z)$ is the Gamma function, and

$$\langle x \rangle_+ = \delta - \frac{1}{\nu} \frac{\Gamma(2/\alpha)}{\Gamma(1/\alpha)} \tag{3.58}$$

Calculating the mean square fluctuation is equally simple. This concludes our survey of well-known one-point densities.

3.6.5 Levy distributions

Mandelbrot (1966) argued that the idea of aggregation should be important in economics, where data are typically inaccurate and may arise from many different underlying causes, as in the growth populations of cities or the number and sizes of firms. He therefore asked which distributions have the property (unfortunately called "stable") that, with the n different densities f_k in the CLT replaced by exactly the same density f, we obtain the same functional form under aggregation, but with different parameters a, where a stands for a collection $(a_1,\ldots,a_m)$ of m parameters including the time variables t_k:

$$\tilde{f}(x,\alpha) = \int \ldots \int dx_1 f(x_1,\alpha_1) \ldots dx_n f(x_n,\alpha_n) \delta(x - \sum x_k/\sqrt{n}) \qquad (3.59)$$

Here, the connection between the aggregate and basic densities is to be given by self-affine scaling

$$\tilde{f}(x) = Cf(\lambda x) \qquad (3.60)$$

As an example, the convolution of any number of Gaussians is again Gaussian, with a different mean and standard deviation than the individual Gaussians under the integral sign. Levy had already answered the more general question, and the required distributions are called Levy distributions. Levy distributions have the fattest tails (the smallest tail exponents). However, in contrast with Mandelbrot's motivation stated above, the Levy distribution does have a well-defined underlying stochastic dynamics, namely, the Levy flight (Hughes et al., 1981).

Denoting the Fourier transform by $f(k)$,

$$f(x) = \int \phi(k) e^{ikxdk} dk \qquad (3.61)$$

the use of (3.61) in the convolution (3.59) yields

$$\tilde{f}(x) = \int dk \Phi(k) e^{ikx} = \int dk \phi^n(k) e^{ikx} \qquad (3.62)$$

so that the scaling condition (3.60) yields

$$\phi^n(k) = C\phi(k/\lambda)/\lambda \qquad (3.63)$$

The most general solution was found by Levy and Khintchine (Gnedenko, 1967) to be

$$\ln \phi(k) = \begin{cases} i\mu k - \gamma |k|^\alpha [1 - i\beta \dfrac{k \tan(\pi\alpha/2)}{|k|}], \alpha \neq 1 \\ i\mu k - \gamma |k|[1 + i\beta \dfrac{2k \ln|k|}{\pi |k|}], \alpha = 1 \end{cases} \quad (3.64)$$

Denote the Levy densities by $L_a(x,\Delta t)$. The parameter β controls asymmetry, $0 < \alpha \leq 2$, and only three cases are known in closed form: $\alpha = 1$ describes the Cauchy distribution,

$$L_1(x, \Delta t) = \frac{1}{\pi \Delta t}\left(\frac{1}{1 + x^2/\Delta t^2}\right) \quad (3.65)$$

$\alpha = 1/2$ is Levy-Smirnov, and $\alpha = 2$ is Gaussian. For $0 < \alpha < 2$ the variance is infinite. For x large in magnitude and $\alpha < 2$ we have

$$L_\alpha(x) \approx \frac{\mu A_\pm^\alpha}{|x|^{1+\alpha}} \quad (3.66)$$

so that the tail exponent is $\mu = 1 + \alpha$. The truncated Levy distribution was applied to a stock index by Mantegna and Stanley (2000).

There is a scaling law for both the density and also the peak of the density at different time intervals that is controlled by the tail exponent α. For the symmetric densities

$$L_\alpha(x, \Delta t) = \frac{1}{\pi}\int_{-\infty}^{\infty} dk e^{ikx - \gamma k^\alpha \Delta t} \quad (3.67)$$

so that

$$L_\alpha(x, \Delta t) = \Delta t^{-1/\alpha} L_\alpha(x/\Delta t^{1/\alpha}, 1) \quad (3.68)$$

A data collapse is predicted with rescaled variable $z = x/\Delta t^{1/\alpha}$. The probability density for zero return, a return to the origin after time Δt, is given by

$$L_\alpha(0, \Delta t) = L_\alpha(0, 1)/\Delta t^{1/\alpha} \quad (3.69)$$

Many interesting properties of Levy distributions are presented in Scalas et al. (2000) and Mainardi et al. (2000).

3.7 Stochastic processes

A random variable x is, by definition, any variable that is described by a probability distribution. Whether a "random variable" evolves deterministically in time via deterministic chaotic differential equations (where *nothing* in the time evolution is random) or "randomly" (via *stochastic* differential equations or sdes) is not implied in the definition. Chaotic differential equations generate pseudo-random time series $x(t)$ and corresponding probability distributions perfectly deterministically. In this text, we are not concerned with deterministic dynamical systems (excepting Chapter 2) because they are not indicated by market data analysis. The reason for this is simple. Deterministic dynamics is smooth at the smallest time scales. There, the motion is equivalent via a local coordinate transformation to constant-velocity motion. Random processes (stochastic processes), in contrast, have unpredictable jumps at even the shortest time scales, as in the stock market over one tick (where typically $\Delta t \approx 1$ second). Hence, in this text we concern ourselves with the methods of the theory of stochastic processes, but treat here only the ideal case of continuous time processes because the discrete case is much harder to handle analytically, and because finance data can be fit to a large extent using continuous time models.

By a stochastic or random *process* we mean one where the random variable $x(t)$ obeys a *stochastic* equation of motion, an equation of motion driven by *noise*. By noise, we will mean a drift-free random variable specified by a probability distribution, where the random variable does not evolve in time deterministically. The discrete random walk or continuous Brownian motion provide the canonical examples. We will call a realization of a stochastic process a random time series.

3.7.1 Introduction to stochastic processes

Before discussing stochastic equations of motion, we ask: how can a random time series $x(t)$ be characterized? According to Kolmogorov, we can define a specific stochastic process precisely if and only if we can specify the complete, infinite hierarchy of joint probability distributions. From an empirical standpoint we can at best obtain finitely many histograms from measurement of N different runs for n different times, using the frequency definition of probability, so we can never hope to specify a process uniquely; the best we can hope for is to specify some class of processes.

Let $P_1(x)$ denote the probability to find a value $X < x$ at time t. This is the one-point distribution of x. Then $P_2(x, t; x', t')$ denotes the probability to find both $X < x$ at time t and $X' < x'$ at time t', and so on up to $P_n(x_1, t_1; \ldots; x_n, t_n)$. Clearly, both the number N of runs and the number n of times that we strobe/observe the system must be large in order to have any hope of getting good statistics (meaning reliable histograms). Statistical independence of events, complete lack of correlations at all levels, means that $P_n(x_1, t_1; \ldots; x_n, t_n) = P_1(x_1, t_1) \ldots P_1(x_n, t_n)$, but this is the rare exception. In order to discuss correlations we will need P_2, at the very least. We can expect in practice that P_n will be ill-defined, experimentally, because we've reached the limit of resolution of our measurements. Moreover, we will discover that for certain common classes of time series it's nontrivial to get enough points from even a long time series to discover P_1. Nonuniqueness in modeling is inherent in the fact that we can only hope to discover the lowest few distributions P_n from empirical data. The class of models that we are able to deduce from data may be unique at the level of P_2, but then will be nonunique at the level of P_3 and beyond. But in that case what cannot be discovered from measurement should not be interpreted as license for invention from mathematical imagination.

Although the above background is necessary for theoretical orientation, we will see in Chapter 7 that densities are very, very hard to obtain empirically. Generally, one must settle in practice for simple averages and pair correlations.

Looking ahead, we introduce the hierarchy of probability densities f_n via

$$dP_n(x_1,t_1;\ldots;x_n,t_n) = f_n(x_1,t_1;\ldots;x_n,t_n)dx_1 \ldots dx_n \qquad (3.70)$$

3.7.2 Conditional probability densities

Correlations cannot be described by a one-point density $f_1(x,t)$. The two-point density $f_2(y,t;x,s)$ is required to calculate pair correlations $\langle x(t)x(s) \rangle = \int dy dx\, yx f_2(y,t;x,s)$. The one-point density suffices if and only if the variables $x(t)$ and $x(s)$, $t \neq s$, are statistically independent, if $f_2(y,t;x,s) = f_1(y,t)f_1(x,s)$. Statistical independence is the rare exception, not the rule. Generally, values of x at two different times are correlated by the underlying dynamics. Our aim is to learn how to deduce and model dynamics faithfully by using adequate empirical data. That will require discovering the pair correlations.

Consider a time series $x(t)$ representing one run of a stochastic process. Empirically, we can only strobe the system a finite number of times, so measurements of $x(t)$ take the form of $\{x(t_k)\}$, $k = 1,\ldots,n$ where n is the number of measurements made. If we can extract good enough histograms from the data, then we can construct the hierarchy of probability densities $f_1(x,t), f_2(x_1,t_1;x_2,t_2), \ldots, f_k(x_1,t_1;\ldots;x_k,t_k)$ where $k \ll n$ (the one-point density f_1 reflects a specific choice of initial condition in data analysis). To get decent histograms for f_n, one would then need a much longer time series.

We note that

$$f_{n-1}(x_1,t_1;\ldots;x_{k-1},t_{k-1};x_{k+1},t_{k+1};\ldots;x_n,t_n) = \int dx_k f_n(x_1,t_1;\ldots;x_n,t_n) \qquad (3.71)$$

so that

$$f_1(x,t) = \int dy f_2(y,s;x,t). \tag{3.72}$$

Densities of all orders are normalized to unity if $\int dx x f_1(x,t) = 1$.

Let $t_1 < \ldots < t_n$. Two-point conditional probability densities p_k, or transition probability densities, are defined by

$$f_2(x_2,t_2;x_1,t_1) = p_2(x_2,t_2|x_1,t_1)f_1(x_1,t_1) \tag{3.73}$$

$$f_3(x_3,t_3;x_2,t_2;x_1,t_1) = p_3(x_3,t_3|x_2,t_2;x_1,t_1)p_2(x_2,t_2|x_1,t_1)f_1(x_1,t_1) \tag{3.74}$$

and more generally as

$$\begin{aligned}f_n(x_n,t_n;\ldots;x_1,t_1) &= p_n(x_n,t_n|x_{n-1},t_{n-1};\ldots;x_1,t_1)\\ f_{n-1}(x_{n-1},t_{n-1};\ldots;x_1,t_1) &= p_n(x_n,t_n|x_{n-1},t_{n-1};\ldots;x_1,t_1)\ldots\\ p_2(x_2,t_2|x_1,t_1)&f_1(x_1,t_1)\end{aligned} \tag{3.75}$$

where p_n is the two-point conditional probability density to find x_n at time t_n, given the last observed point (x_{n-1},t_{n-1}) and the previous history $(x_{n-2},t_{n-2};\ldots; x_1,t_1)$. The previous history can be regarded as discrete, finite memory. There are also processes like fractional Brownian motion with uncountable memory encoded in f_2 via the pair correlations.

A Markov process is a process with no memory ("no after effect") other than that of the last observed point (x_{n-1},t_{n-1}). This yields

$$f_n(x_n,t_n;\ldots;x_1,t_1) = p_2(x_n,t_n|x_{n-1},t_{n-1})\ldots p_2(x_2,t_2|x_1,t_1)f_1(x_1,t_1) \tag{3.76}$$

because

$$p_k(x_k,t_k|x_{k-1},t_{k-1};\ldots;x_1,t_1) = p_2(x_k,t_k|x_{k-1},t_{k-1}) \tag{3.77}$$

for $k = 3,4,\ldots$ so p_2 cannot depend on an initial state (x_1,t_1) or on any previous state other than the last observed point (x_{k-1},t_{k-1}). By "memory" we mean history other than the last observed state; a Markov process by definition has no memory. Only in the absence of memory does the two-point density p_2 describe the complete time evolution of the dynamical system. The Markov process is a dynamically interesting generalization of the less useful notion of statistical independence of the n random variables x_k, whereby

$$f_n(x_n,t_n;\ldots;x_1,t_1) = f_n(x_n,t_n)\ldots f_2(x_2,t_2)f_1(x_1,t_1) \tag{3.78}$$

For an arbitrary process with or without memory it follows that

3.7 Stochastic processes

$$p_{k-1}(x_k,t_k|x_{k-2},t_{k-2};\ldots;x_1,t_1) = \int dx_{k-1} p_k(x_k,t_k|x_{k-1},t_{k-1};\ldots;x_1,t_1) \\ p_{k-1}(x_{k-1},t_{k-1}|x_{k-2},t_{k-2};\ldots;x_1,t_1) \quad (3.79)$$

so that

$$p_2(x_3,t_3|x_1,t_1) = \int dx_2 p_3(x_3,t_3|x_2,t_2;x_1,t_1) p_2(x_2,t_2|x_1,t_1). \quad (3.80)$$

Normalization of conditional densities follows easily: from

$$\int dy dx f_2(y,s;x,t) = \int dy f_1(x,t) = 1 \quad (3.81)$$

we obtain

$$\int dy p_2(y,s|x,t) = 1 \quad (3.82)$$

which also reflects conservation of probability. From the definition of conditional probability, where x was observed to have occurred at time t, we obtain

$$p_2(y,t|x,t) = \delta(y-x) \quad (3.83)$$

For a Markov process we have $p_n = p_2$ for $n = 2,3,\ldots$ so that

$$p_2(x_3,t_3|x_1,t_1) = \int dx_2 p_2(x_3,t_3|x_2,t_2) p_2(x_2,t_2|x_1,t_1) \quad (3.84)$$

The Markov property is expressed by $p_n = p_2$ for all $n \geq 3$, the complete lack of history-dependence excepting the last observed point. The Chapman-Kolmogorov (CK) equation (3.84) is a necessary but insufficient condition for a Markov process. A CK equation (3.84) does not imply a Markov process (McCauley, 2008b).

The unconditioned average over initial conditions (the initial condition is defined by specifying $f_1(x,t_0)$) is an average over all variables, for example

$$\langle x(t)x(s)\rangle = \int dy dx xy p_2(y,s|x,t) f_1(x,t) \quad (3.85)$$

Unconditioned averages must be distinguished from conditional averages, where in the latter a specific "last position" was observed, for example

$$\langle x(t)\rangle_{cond} = \int dy y p_2(y,t|x,s) \quad (3.86)$$

so that

$$\langle x(t)x(s)\rangle = \int dxx \langle x(t)\rangle_{\text{cond}} f_1(x,s) \tag{3.87}$$

includes an average over all possible initial positions. In finance theory (e.g. option pricing and in some definitions of volatility) we often meet conditional averages. For example, the price is always known at the present time t, so averages over prices at future times $t + T$ can in principle be conditioned on the fact that we know $p(t)$. In general, $f_1(x,t)$ reflects the free choice of initial condition $f_1(x,t_0)$ rather than the dynamics.

3.7.3 Martingales

A Martingale, more precisely, a "local Martingale," is defined by the condition

$$\langle x(t)\rangle_{\text{cond}} = \int dyy p_2(y,t|x,s) = x \tag{3.88}$$

and generalizes the idea of a fair game to continuous time processes. The idea of a fair game is that there is no gain or loss. The Martingale condition states that the expected future average equals the last observed value and there's no trend to move you away from where you stand right now on the average, so that the expected value of your later net worth is what you hold at the moment.

A Martingale has no trend, $d\langle x\rangle/dt = 0$, because

$$\langle x(t+T)\rangle_{\text{cond}} = x(t) \tag{3.89}$$

where the average is conditioned on having observed the point x at an earlier time t,

$$\langle x(t+T)\rangle_{\text{cond}} = \int dyy p_2(y,t+T|x,t) = x \tag{3.90}$$

Using this condition in the unconditioned average (hereafter simply called "the average")

$$\langle x(t)x(s)\rangle = \int dy dx yx p_2(y,t|x,s) f_1(x,t) \tag{3.91}$$

yields a powerful result:

$$\langle x(t)x(s)\rangle = \langle x^2(s)\rangle, s < t \tag{3.92}$$

We can reverse the argument to show, since f_1 arises from a freely chosen initial condition, that (3.91) implies a Martingale. We can take (3.92) as the Martingale condition.

We show next that the condition (3.92) rules out memory (history dependence) in a Martingale at the level of simple averages and pair correlations. From (3.92) we derive the more general condition that there is no correlation in increments/displacements occurring in nonoverlapping time intervals; the increment autocorrelation

$$\langle (x(t) - x(t-T))(x(t+T) - x(t)) \rangle = \int dxdydz(z-y)(y-x)$$
$$p_2(z, t+T|y, t; x, t-T) p_2(y, t|x, t-T) f_1(x, t-T) = 0 \quad (3.93)$$

vanishes via term-by-term cancellation. This means that nothing that happened in the past can be used to predict future patterns of behavior *at the level of pair correlations*. We can also take (3.93) as the Martingale condition, but there may be memory in a Martingale.

Memory, history dependence, can appear in a Martingale in a subtle way. We now follow Hänggi and Thomas (1977) to explain in part what we mean by "subtle." In an obvious shorthand notation, starting with $f_2(x_3; x_2) = \int f_3(x_3; x_2; x_1) dx_1$ and using $f_2(x_3; x_2) = p_2(x_3|x_2) f_1(x_2) = p_2(x_3|x_2) \int p_2(x_2|x_1) f_1(x_1) dx_1$ we obtain

$$p_2(x_3, t_3|x_2, t_2) = \frac{\int dx_1 p_3(x_3, t_3|x_2, t_2; x_1, t_1) p_2(x_2, t_2|x_1, t_1) f_1(x_1, t_1) dx_1}{\int p_2(x_2, t_2|x_1, t_1) f_1(x_1, t_1) dx_1} \quad (3.94)$$

which is a functional of the initial state $f_1(x_1, t_1)$ in which the system was prepared at the initial time t_1 unless the process is Markovian. In a non-Markov system one may sometimes be able to mask this dependence on state preparation by choosing the initial condition to be $f_1(x_1, t_1) = \delta(x_1)$. If, instead, we would or could choose $f_1(x_1, t_1) = \delta(x_1 - x'_0)$ at $t_1 = 0$, for example, then we would obtain $p_2(x_3, t_3|x_2, t_2) = p_3(x_3, t_3|x_2, t_2; x'_0)$. So in this case, what appears superficially as p_2 is really a special case of p_3. A Martingale may have memory of a discrete set of states $(x_{n-1}, \ldots, x_1)$ in the past, but cannot have memory of an entire continuous past trajectory $x(t)$. The latter sort of memory (exemplified by fractional Brownian motion in Chapter 6) violates (3.94) because the pair correlations violate (3.92) and reflect the strong memory.

One can introduce a trend by adding a drift term to a Martingale. There is an intimate connection between Martingales and detrendable processes, which are called Ito processes. An Ito process is quite simply a trend plus a Martingale, where the Martingale may be understood as noise, albeit generally not white noise. The reason that a Martingale is a form of noise is simply that it doesn't transport you anywhere on the average.

There are no elementary or easily readable references on Martingales, but see Steele (2000) and especially both books by Durrett (1984, 1996). Certain elementary exercises are assigned in those texts, and the reader can get a better feeling for Martingales by working them. Durrett is hard to read, his notation at one stage is not clearly defined and must be decoded by the reader, but many sections can be skipped and the books are worth the effort.

3.7.4 Detrending a Martingale plus drift

We can classify processes according to those that can be detrended and those that can't. The first class, we can think of as drift plus noise or, more precisely, drift plus a Martingale. Martingales can be understood as noise sources with no increment autocorrelations: with increments $M(t,T) = M(t+T) - M(t)$ and $M(t,-T) = M(t) - M(t-T)$, we have $\langle M(t,T)M(t,-T)\rangle = 0$. This leads us to study Ito processes, which include Martingales, Markov processes, and generalizations of Markov processes with drift, and also Ito processes with finite memory (the latter are not treated here). We must first develop the connection between Martingales and the absence of trend, and then show how trend can be added.

Consider a Martingale $M(t)$, with $M(t_0) = 0$, plus a drift $A(t)$. We suggest that the general form of a detrendable process is

$$x(t) = x(t_0) + A(t) + M(t) \tag{3.95}$$

so that $\langle x(t) \rangle_{\text{cond}} = A(t)$, and $M(t)$ describes the noise. The next point is to show that the drift term can be written as $A(t) = \int R(x(s),s)ds$, a path-dependent functional of the stochastic process x. First, we must define the drift coefficient R.

Whenever it exists the drift coefficient is defined by

$$R(x,t) \approx \frac{1}{T} \int_{-\infty}^{\infty} dy(y-x)p_2(y,t;x,t-T) \tag{3.96}$$

as T vanishes. Here is our main point: if $R = 0$ then we obtain from (3.96) the condition

$$\int_{-\infty}^{\infty} dy\, y p_2(y,t;x,t-T) = x \tag{3.97}$$

so that the average at a later time is given by the last observed point in the time series, $\langle x(t+T)\rangle_{cond} = x(t)$. This is the notion of a fair game: there is no change in the average of x as t increases.

Using the delta function initial condition for p_2 as T vanishes, we can generalize

$$\langle x(t)\rangle_{cond} \approx x(t_0) + R(x(t),t)T \tag{3.98}$$

valid for small $T = t - t_0$ to obtain

$$\langle x(t)\rangle_{cond} = x(t_0) + \int_{t_0}^{t} \langle R(x(s),s)\rangle_{cond} ds \tag{3.99}$$

which is valid for all t. The trend is described by the average drift

$$d\langle x(t)\rangle_{cond}/dt = \langle R\rangle_{cond} \tag{3.100}$$

This suggests a stochastic process of the form

$$x(t) = x(t_0) + \int_{t_0}^{t} R(x(s),s)ds + M(t) \tag{3.101}$$

where, as we will see, the Martingale M may also depend on the path $x(t)$. The general specification of $M(t)$ using the most basic Martingale can be stated only after we prove Ito's theorem.

Our main point is that the possibility of detrending by subtracting a well-defined average drift presumes that the irreducible underlying noise source is a Martingale. So we could divide stochastic processes into those that satisfy the Martingale condition $\langle x(t)\rangle_{cond} = x(t_0)$ and those that don't. This classification is too broad to be very useful, as the latter class includes both fractional Brownian motion and Markov processes, which are very different at the level of pair correlations and beyond.

For a time series describing a Martingale, the problem of forecasting is trivial. Given any set of n points in a time series, $\{x(t_k)\}$, $k = n, n-1, \ldots, 2, 1$, where $t_n > t_{n-1} > \ldots > t_2 > t_1$, and the hierarchy of transition densities p_n, if the increments are uncorrelated then the best systematic forecast of the future is the conditional average $\langle x(t_k)\rangle = x(t_{k-1})$. That is, the future is forecast on the average by the last observed point in the time series,

$$\int dx_n x_n p_n(x_n, t_n | x_{n-1}, t_{n-1}; \ldots; x_1, t_1) = x_{n-1} \tag{3.102}$$

All previous observations $(x_{n-1}, \ldots, x_1)$ don't contribute. This is nontrivial precisely because Martingales can admit history dependence. The point is that at the level of simple averages and pair correlations the history dependence cannot be detected. The simplest example of a Martingale is a drift-free Markov process.

In order to appreciate that economic and finance data are strongly nonstationary, we exhibit the idea of a stationary process for comparison. Toward that end, we generalize the idea of equilibrium to stochastic processes. *Statistical equilibrium* is a property of a *stationary* stochastic process. A nonstationary stochastic process is not asymptotically stationary; it does not approach statistical equilibrium. To make matters worse for the reader, these ideas should not be confused with stationary and nonstationary increments!

3.7.5 Stationary vs nonstationary processes

Time-translational invariance means that every density in the infinite hierarchy is *invariant* under a shift of time origin,

$$f_n(x_1, t_1 + T; \ldots; x_n, t_n + T) = f_n(x_1, t_1; \ldots; x_n, t_n) \tag{3.103}$$

This means that the *normalizable* one-point density, *if it exists*, must be t-independent, i.e. $f_1(x,t) = f_1(x)$. The condition is important because we'll exhibit Markov processes where time-translational invariance holds for f_n, $n > 2$, but where there is no normalizable one-point density f_1. Such processes are nonstationary; the normalizable one-point t-independent density of a stationary process describes statistical equilibrium.

A *time-translationally invariant* Markov process defines a one-parameter semi-group $U(t_2, t_1)$ of transformations (A. Friedman, 1975), where $p_2(x_n, t_n | x_{n-1}, t_{n-1}) = p_2(x_n, t_n - t_{n-1} | x_{n-1}, 0)$. Clearly, the drift and diffusion coefficients are t-independent here. An arbitrary time-translationally invariant Markov process generally does not possess a normalizable stationary one-point density, and therefore is not a stationary process. This will be illustrated below via an example.

Weak stationarity or "wide sense" stationarity (Wax, 1954; Yaglom and Yaglom, 1962; Stratonovich, 1963; Gnedenko, 1967) requires only that the normalizable densities f_1 and f_2 are time-translationally invariant, so that the mean and variance are constants and the pair correlations are stationary, $\langle x(t)x(t+T) \rangle = \langle x(0)x(T) \rangle$. Weak stationarity is adequate for defining statistical equilibrium and was introduced from the practical standpoint: densities (histograms) generally cannot be obtained from time series due to sparseness of data; simple averages and pair correlations often can be measured quite accurately. The difficulty of extracting densities empirically will be addressed and illustrated in Chapter 7 for nonstationary processes.

A stationary process describes a system in or near statistical equilibrium, or a driven steady state. Because the one-point density is t-independent, all moments are constants. *In particular, the average and variance are constants.* By an asymptotically stationary process we mean that $f_1(x,t)$ approaches arbitrarily closely to a t-independent density $f_1(x)$ as t becomes large, and that the pair correlations become independent of the starting time t.

Next, we introduce the important ideas of stationary and nonstationary increments. As an overview, nonstationary increments are restricted to nonstationary processes, but stationary increments may also occur in a nonstationary process. All this will be illustrated below.

3.7.6 Nonstationary increments

Stationary increments $x(t,T)$ of a nonstationary process $x(t)$ are defined by

$$x(t,T) = x(t+T) - x(t) = x(0,T) \qquad (3.104)$$

"in distribution," and by nonstationary increments (Stratonovich, 1963) we mean that

$$x(t+T) - x(t) \neq x(0,T) \qquad (3.105)$$

For stationary increments (Mandelbrot and van Ness, 1968; Embrechts and Maejima, 2002), time-translational invariance of the one-point density and transition density (weak stationarity) will be seen below to be sufficient but unnecessary.

To explain the meaning of "equality in distribution," let $z = x(t,T)$ denote the increment, or difference. Then the one-point increment density is given by

$$f(z,t,t+T) = \int f_2(y,t+T;x,t)\delta(z-y+x)\mathrm{d}x\mathrm{d}y \qquad (3.106)$$

which yields

$$f(z,t,t+T) = \int p_2(x+z,t+T|x,t)f_1(x,t)\mathrm{d}x \qquad (3.107)$$

For stationary increments this one-point density must be independent of the starting time t and can depend only on (z,T), yielding $f(x,t,t+T) = f(z,0,T)$, whereas in the nonstationary increment case the t-dependence of f remains.

We showed above that Martingale increments are uncorrelated,

$$\langle x(t,T)x(t,-T) \rangle = 0 \qquad (3.108)$$

Combining

$$\left\langle (x(t+T)-x(t))^2 \right\rangle = \left\langle x^2(t+T) \right\rangle + \left\langle x^2(t) \right\rangle - 2\langle x(t+T)x(t)\rangle \quad (3.109)$$

with (3.92) we get

$$\left\langle (x(t+T)-x(t))^2 \right\rangle = \left\langle x^2(t+T) \right\rangle - \left\langle x^2(t) \right\rangle \quad (3.110)$$

which depends on both t and T, excepting the special case where the variance $\langle x^2(t) \rangle$ is linear in t. *Martingale increments are uncorrelated and are generally nonstationary*. This result is of central importance for finance, and for making the so-called efficient market hypothesis precise.

3.7.7 Stationary increments

We've shown that, at the pair correlation level of description, a Martingale cannot be distinguished from a drift-free Markov process.

Consider next the class of stochastic processes with stationary increments: $x(t,T) = x(0,T)$ "in distribution." The class is wide because it's dynamically nonselective: both stationary and nonstationary processes are included, both efficient and nonefficient markets are included. Following common practice, we will sometimes write $x(0,T) = x(T)$, but there we must avoid the nonsensical misinterpretation that "$\langle x(0,T)x(0,-T)\rangle = -\langle x^2(T)\rangle = 0$" for the case where the increments are stationary and the variance is linear in the time. In the context of increments, by the notation "$x(T)$" we will always mean $x(0,T)$.

We begin simply with

$$-2\langle x(t+T)x(t)\rangle = \left\langle (x(t+T)-x(t))^2 \right\rangle - \left\langle x^2(t+T) \right\rangle - \left\langle x^2(t) \right\rangle, \quad (3.111)$$

then using increment stationarity on the right hand side of (3.111) we obtain

$$-2\langle x(t+T)x(t)\rangle = \left\langle x^2(T) \right\rangle - \left\langle x^2(t+T) \right\rangle - \left\langle x^2(t) \right\rangle \quad (3.112)$$

which differs significantly from (3.92). For increments with nonoverlapping time intervals, the simplest autocorrelation function is

$$\begin{aligned} 2\langle (x(t)-x(t-T))(x(t+T)-x(t))\rangle \\ = \left\langle (x(t+T)-x(t-T))^2 \right\rangle \\ - \left\langle (x(t)-x(t-T))^2 \right\rangle - \left\langle (x(t+T)-x(t))^2 \right\rangle \\ = \left\langle x^2(2T) \right\rangle - 2\langle x^2(T)\rangle \end{aligned} \quad (3.113)$$

which generally does not vanish. *Stationary increments of a nonstationary process are strongly correlated if the process variance is nonlinear in the time t.* And by process variance we mean $\sigma^2(t) = \langle x^2(t) \rangle$ measured from $x(0) = 0$. In the literature the mean square fluctuation $\langle x^2(t,T) \rangle$ about the point $x(t)$ is too often called "the variance." This should instead be denoted as "the increment variance." Precision of language in mathematics is absolutely necessary, otherwise confusion rather than understanding is generated.

In economics, it's quite generally believed that increments are stationary and that increment stationarity implies ergodicity. We next show that the latter cannot hold except for two exceptional processes, processes that don't appear as noise in real economic data.

3.7.8 Increment stationarity vs ergodicity

As is often assumed in economics, let $T =$ one period in a nonstationary model with stationary increments. Then $x(0,1)$ has a well-defined stationary density $f(z,t,t+1) = f(z,0,1)$, but there generally is no ergodicity accompanying increment stationarity. There is one exception. Consider a time- and space-translationally invariant drift-free Markov process. For $t_n - t_{n-1} = \ldots = t_1 - t_0 = T$, the Markov condition then yields the density f_{n+1} as $p_2(x_n - x_{n-1}, 1|0,0) p_2(x_{n-1} - x_{n-2}, T|0,0) \ldots p_2(x_2 - x_1, T|0,0) p_2(x_1 - x_0, T|0,0)$ with $f_1(x,t) = p_2(x_1 - x_0, T|0,0)$. With T taken as time variable the increment process is nonstationary as T increases, but with T fixed ($T = 1$, for example) the increments are iid (stationary, statistically independent, and identically distributed). In this case the Tschebychev inequality (the law of large numbers) guarantees ergodicity: time averages of increments will converge with probability one to ensemble averages. We'll see below that the increment process so described must be Gaussian with variance linear in T, and is in fact the so-called Wiener process, the Wiener process being the only (x,t)-translationally invariant Martingale, and this assertion is easily proven below using the Fokker–Planck pde.

3.7.9 Gaussian processes

In order to introduce the fundamental stochastic process out of which all others can be built via "stochastic integration," the Wiener process $B(t)$, we first define Gaussian stochastic processes. This class includes Markov processes and Martingales, but is much larger. It also includes fractional Brownian motion (Chapter 6), and other strong memory processes of interest primarily in near-equilibrium statistical physics based on an assumption of correlated noise. In other words, Gaussian processes include very many dynamically unrelated processes.

A Gaussian process has the *n*-point densities (see Wang and Uhlenbeck in Wax, 1954)

$$f_n(x_1, t_1; \ldots; x_n, t_n) = \frac{1}{(2\pi)^{n/2} \det B^{1/2}} e^{-(x-\langle x \rangle)^\dagger B^{-1}(x-\langle x \rangle)/2} \tag{3.114}$$

where the matrix B is defined by the pair correlations,

$$B_{kl} = \langle x_k x_l \rangle \tag{3.115}$$

That is, a Gaussian process is completely defined by its mean and pair correlations.

The two-point and one-point densities have the general form

$$f_2(x, t_1; y, s) = \frac{1}{2\pi\sigma(t)\sigma(s)(1-\rho^2)^{1/2}} e^{-(x^2/\sigma^2(t) + y^2/\sigma^2(s) - 2\rho xy/\sigma(t)\sigma(s))/2(1-\rho^2)} \tag{3.116}$$

and

$$f_1(y, s) = \frac{1}{\sqrt{2\pi\sigma^2(s)}} e^{-(x-a(s))^2/2\sigma^2(s)} \tag{3.117}$$

Using the notation in the papers by Harry Thomas and his students, the conditional density $p_2 = f_2/f_1$ is given by

$$p_2(x, t|y, s) = \frac{1}{\sqrt{2\pi K(t,s)}} e^{-(x-m(t,s)y-g(t,s))^2/2K(t,s)} \tag{3.118}$$

From (3.118) follows

$$\langle x(t) \rangle_{\text{cond}} = \int_{-\infty}^{\infty} dx\, x\, p_2(x, t|y, s) = m(t, s)y + g(t, s) \tag{3.119}$$

so that

$$\langle x(t)x(s) \rangle = m(t, s)(\sigma^2(s) + a^2(s)) + g(t, s) \tag{3.120}$$

The time evolution of the one-point density is given by

$$\begin{aligned} a(t) &= g(t, s) + m(t, s)a(s) \\ \sigma^2(t) &= K(t, s) + m^2(t, s)\sigma^2(s) \end{aligned} \tag{3.121}$$

These processes are quite generally non-Markovian. For a Martingale, we need $m(t,s) = 1$ with $g(t,s) = 0$, and $\sigma^2(t)$ generally depends on t.

3.8 Stochastic calculus

The conditions such that the transition density p_2 satisfies the CK equation (Hänggi et al., 1978) are (3.84)

$$\begin{aligned} m(t,t_0) &= m(t,s)m(s,t_0) \\ g(t,t_0) &= g(t,s) + m(t,s)g(s,t_0) \\ K(t,t_0) &= K(t,s) + m^2(t,s)K(s,t_0) \end{aligned} \quad (3.122)$$

A Gaussian Markov process must satisfy these relations, but a Gaussian process satisfying these relations is not necessarily Markovian. Simple examples are given in McCauley (2008b).

3.7.10 The Wiener process

The Wiener process $B(t)$ is defined as a Gaussian process with statistically independent, stationary increments $B(t,T) = B(0,T)$. We can apparently choose the initial condition $B(0) = 0$ at $t_0 = 0$, so that $B(t,0) = B(t) - B(0) = B(t)$, with stationary increments $B(t,T) = B(0,T) = $ "$B(T)$" in distribution. Statistical independence of the increments implies vanishing increment autocorrelations, $\langle B(0,T)B(0,-T)\rangle = 0$, which requires in turn that the variance is linear in t, $\langle B^2(t)\rangle = t$. The Wiener process therefore has the transition density

$$p_2(B,t|B_0,t_0) = \frac{1}{\sqrt{2\pi(t-t_0)}} e^{-(B-B_0)^2/2(t-t_0)} \quad (3.123)$$

and is a time-translationally invariant Martingale. It is also the simplest time-translationally invariant Markov process. There is no corresponding normalizable time-independent one-point density; the Wiener process is fundamentally nonstationary (the variance increases with t). As one-point density for calculating averages like the variance, we use $f_1(B,t) = p_2(B,t|0,0)$.

By "white noise" in statistical physics and radio theory (albeit *not* in econometrics) is meant $\eta = dB/dt$. This pointwise undefined derivative makes sense in a certain statistical sense (see Wax (1954) or Stratonovich (1963)), where one can show that η is formally a stationary process. In this text we will work only with the Wiener process via Ito calculus, and not with white noise. The Wiener process is the simplest example of a nonstationary process. As we've pointed out above, the Wiener process is also singular as a stationary increment process because the Markov condition for fixed lag times yields iid increments.

3.8 Stochastic calculus

The motivation for stochastic calculus is as follows. Without noise, a financial instrument (savings deposit, certificate of deposit (CD), or money market

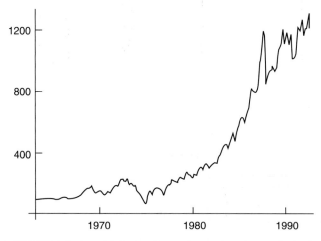

Figure 3.1 UK FTA index, 1963–92. From Baxter and Rennie (1995), fig. 3.1.

deposit) simply pays an interest rate μ, so that the price obeys $dp = \mu p dt$. A bond pays a definite interest rate but also fluctuates in price according to interest rate competition with the central bank, so noise must be added. For a bond, we have $dp = \mu p\, dt + \textit{noise}$, but what sort of noise? If we would take "noise" $= \sigma_1 dB$, where B is the Wiener process, then we would obtain the stationary predictions of the Ornstein–Uhlenbeck (OU) model (described below). But stationarity is not a property of financial markets. The next simplest assumption is that $dp = \mu p dt + p\sigma_1 dB$, which is the lognormal pricing model (also described below) and is nonstationary. Although we've argued from the standpoint of bonds, this equation has also been applied historically to stock and foreign exchange markets. Figure 3.1 shows a stock index, and Figure 3.2 shows the prediction of the lognormal pricing model (which we can solve once we know stochastic calculus). Although the details are wrong, one sees visually that the model has some correct qualitative features. The point is to develop stochastic calculus in order to introduce the empirically correct class of model.

3.8.1 Ito's theorem

Stochastic calculus is developed based on the Wiener process. We begin by introducing the stochastic differential

$$dy = R(B,t)dt + b(B,t)dB \qquad (3.124)$$

in combination with the stochastic integral

3.8 Stochastic calculus

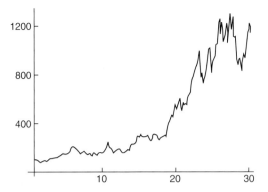

Figure 3.2 Exponential Brownian motion dp = Rpdt + σpdB with constant R and σ. Baxter and Rennie (1995), fig. 3.6.

$$y(t) = y(t_0) + \int_{t_0}^{t} R(B(s), s) ds + \int_{t_0}^{t} b(B(s), s) dB(s) \qquad (3.125)$$

The first term is an ordinary integral of a random variable $B(t)$, but the second term is a "stochastic integal" and is denoted by the symbol for the "Ito product"

$$b \bullet \Delta B = \int_{0}^{t} b(B(s), s) dB(s) \qquad (3.126)$$

which is strictly defined by

$$b \bullet \Delta B = \int_{0}^{t} b(B(s), s) dB(s) \approx \sum_{k=1}^{N} b(B_{k-1}, t_{k-1}) \Delta B_k \qquad (3.127)$$

for large N, where $\Delta B_k = B(t_k) - B(t_{k-1})$.

The main point is that Ito defined the stochastic integral so that the noise does not renormalize the drift (Stratonovich's definition of the stochastic integral is different). Because the integrand is defined in the sum at the *left* end point t_{k-1}, we have

$$\langle b \bullet \Delta B \rangle = 0 \qquad (3.128)$$

because the increments are uncorrelated: $\langle \Delta B_{k-1} \Delta B_k \rangle = 0$ because B_{k-1} occurs before ΔB_k. Functions $b(B,t)$ that satisfy this condition are called "nonanticipating functions," meaning functions determined by the Wiener process at times earlier than time t.

The next order of business is to prove Ito's theorem, and then Ito's lemma. To prove Ito's theorem, we must study the variable $dB(t)^2 = (dB(t))^2$. We must be careful to distinguish $(dB)^2$ from $d(B^2(t))$ in all that follows. We aim to prove that the former behaves deterministically, that in Ito calculus we can use $(dB)^2 = dt$.

To derive Ito's theorem we study the stochastic integral

$$I = \int_0^t dB^2 \approx \sum_{k=1}^N \Delta B_k^2 \qquad (3.129)$$

The variance of I is

$$\sigma_I^2 = N(\langle \Delta B^4 \rangle - \langle \Delta B^2 \rangle^2) \qquad (3.130)$$

Denoting the increment as $\Delta B = B(t + \Delta t) - B(t)$ and using the Gaussian distribution, we get $\langle \Delta B^4 \rangle = 3\Delta t^2$, so that with $\Delta t = t/N$ we get $\sigma_I^2 = 2t^2/N$, which vanishes as N goes to infinity. That is, the width of the density of the random variable $(\Delta B)^2$ vanishes as Δt vanishes, so that density of the random variable $(dB)^2$ is a delta function. This means that $(dB)^2$ is, with probability unity, deterministic and so we can take

$$(dB)^2 = dt \qquad (3.131)$$

"in probability." An immediate consequence of this is that, with $R = 0$, the process variance is

$$\langle y^2(t) \rangle_{cond} = y^2(t_0) + \int_{t_0}^t \langle b^2(B(s), s) \rangle_{cond} ds \qquad (3.132)$$

That is, variances can be easily formulated as ordinary integrals by using Ito calculus. We will next prove Ito's lemma, which is the practical tool needed to do stochastic calculus.

The beauty and usefulness of Ito calculus is twofold. First, in the Ito stochastic integral the noise does not renormalize the drift. The latter effect is built into the definition of the Ito integral, as we've just shown: $\langle b \bullet \Delta B \rangle = 0$, whether the average is conditioned or not. Second, as we show next, given a specific stochastic process $x(t)$, Ito's lemma allows us easily to construct coordinate transformations to define topologically related stochastic processes $y(t)$. Two different processes are topologically equivalent if connected by a continuous, invertible transformation. Ito processes in addition require twice-differentiability of the transformation.

3.8.2 Ito's lemma

Given the Wiener process $B(t)$, we consider the class of twice-differentiable functions $G(B,t)$, which we can understand as a coordinate transformation from the random variable B to define a new random variable $y = G(B,t)$. If the transformation is also invertible then the processes are topologically equivalent. Ito's lemma uses the lowest-order terms in a Taylor expansion

$$dG = \frac{\partial G}{\partial t}dt + \frac{\partial G}{\partial B}dB + \frac{1}{2}\frac{\partial^2 G}{\partial B^2}dB^2 \tag{3.133}$$

to construct the stochastic differential defining the new random variable y:

$$dG = \left(\frac{\partial G}{\partial t} + \frac{1}{2}\frac{\partial^2 G}{\partial B^2}\right)dt + \frac{\partial G}{\partial B}dB \tag{3.134}$$

Note that the first term on the right-hand side is the drift and the second term is the noise term, so that the noise renormalizes the drift in the coordinate transformation.

Using (3.134) we can obtain $y = G(B,t)$ via stochastic integration

$$y = G(B,t) = G(0,0) + \int (\partial G/\partial t + \frac{1}{2}\partial^2 G/\partial B^2)dt + \int \partial G/\partial B dB \tag{3.135}$$

One can use Ito's lemma to evaluate stochastic integrals, or to reduce new stochastic integrals to the evaluation of other ones. Here are two easy examples.

First, consider $y = B^2 - t$. Ito's lemma yields

$$dy = 2BdB \tag{3.136}$$

Conbining this with y we obtain

$$\int_0^B BdB = (B^2 - t)/2 \tag{3.137}$$

Second, consider $G(B,t) = B^3/3 - Bt$. Then

$$dy = (B^2 - t)dB \tag{3.138}$$

so that

$$\int_0^B B^2 dB - \int_0^t sdB(s) = B^3/3 - t \tag{3.139}$$

Note next that the examples y(t) considered have no drift (no "trend"), $\langle y(t) \rangle = G(0,0) = 0$. We have two explicit examples of Martingales.

3.8.3 Local vs global descriptions of stochastic dynamics

Note that the stochastic differential dy provides a "local" description of a stochastic process, one valid in a very small neighborhood of a point (B,t),

$$\Delta y \approx R(B,t)\Delta t + b(B,t)\Delta B \qquad (3.140)$$

The full transformation $y = G(B,t)$ provides a global description of the same process, a description valid over finite (if not infinite) intervals in both B and t.

Let $R = 0$ here. Note further that locally, with $\Delta y = y(t + \Delta t) - y(t)$, the mean square fluctuation is

$$\langle \Delta y^2 \rangle \approx \langle b^2(B,t) \rangle \Delta t \qquad (3.141)$$

so that $b^2(B(t),t)$ is the diffusion coefficient for the transformed Wiener process: the new process y is, by a twice-differentiable coordinate transformation, topologically equivalent to a Wiener process. The stochastic processes important for finance markets are *not* topologically equivalent to Wiener processes, and we will show how to build topologically inequivalent processes from a Wiener process via a stochastic integral equation.

Note that the mean square fluctuation $\langle y(t,T)^2 \rangle$ about an arbitrarily chosen point y(t) is not the process variance. The variance of the process y(t) is defined by $\langle y(t)^2 \rangle$. Confusing the two can arise from insufficient attention to the question of stationary vs nonstationary increments.

3.8.4 Martingales for beginners

The defining condition for a local Martingale M(t) with $M(0) = 0$ is $\langle M(t) \rangle_{cond} = M(t_0)$. This implies $\langle M(t)M(s) \rangle_{cond} = \langle M^2(s) \rangle$ if $s < t$, conditions satisfied by the Wiener process. Note that

$$M(t) = M(t_0) + \int_{t_0}^{t} b(B(s),s)\mathrm{d}B(s) \qquad (3.142)$$

is also a Martingale for nonanticipating functions b(B,t).

We can use transformations $y = G(B,t)$ on Wiener processes to construct Martingales, because a Martingale is generated by any drift-free Ito stochastic process

$$y = G(B,t) = G(0,0) + \int \partial G/\partial B \, dB \tag{3.143}$$

where

$$\langle G(B,t) \rangle = G(0,0) \tag{3.144}$$

It follows by Ito's lemma that

$$dy = (\partial G/\partial t + \frac{1}{2}\partial^2 G/\partial B^2)dt + \frac{\partial G}{\partial B} dB \tag{3.145}$$

We can define infinitely many different Martingales $y = G(B,t)$ by solving the simplest backward-time-diffusive pde

$$\partial G/\partial t + \frac{1}{2}\partial^2 G/\partial B^2 = 0 \tag{3.146}$$

subject to boundary or initial conditions.

With a Martingale $y = G(B,t)$

$$dG = (\partial G/\partial x)dB \tag{3.147}$$

$$\Delta y = G' \bullet dB \tag{3.148}$$

then the additional condition

$$\sigma^2 = \left\langle (\Delta y)^2 \right\rangle = \int \langle G'^2 \rangle dt < \infty \tag{3.149}$$

is required for a global Martingale where the function $D(B,t) = \sqrt{G'(B,t)}$ is identified as the diffusion coefficient.

We're now prepared to study stochastic differential and stochastic integral equations.

3.9 Ito processes

Ito processes are diffusive processes with uncorrelated noise. The latter follows from Ito's theorem in the definition of the stochastic integral. We now illustrate the diffusive property.

3.9.1 Stochastic differential equations

We argued earlier that the general form of a process made up of noise with uncorrelated increments plus a drift term is

$$x(t) = A(t) + M(t) \tag{3.150}$$

where the drift term is given by $A(t) = \int R(x(s),s)ds$ with

$$R(x,t) \approx \frac{1}{T} \int_{-\infty}^{\infty} dy (y-x) p_2(y,t;x,t-T) \tag{3.151}$$

as T vanishes, and $M(t)$ is a Martingale with $M(0) = 0$. Here is the central question for finance theory: what is the most general Martingale that we can write down? Note that the stochastic integral equation

$$x(t) = x(t_0) + \int_{t_0}^{t} b(x(s), s) dB(s) \tag{3.152}$$

is a Martingale so long as $b(x,t)$ is a nonanticipating function. This is the most general form of a Martingale, and appears in the math literature as the Martingale Representation Theorem.

The diffusion coefficient is defined analogous to the drift coefficient by

$$D(x,t) \approx \frac{1}{T} \int_{t}^{t+T} (y-x)^2 p_2(y, t+T; x, t) \tag{3.153}$$

as T vanishes. Note that (without drift)

$$\langle x^2(t+T) \rangle \approx x^2(t) + D(x,t) T \tag{3.154}$$

where $b^2(x,t) = D(x,t)$ is the diffusion coefficient. For a Martingale, the mean square fluctuation about the point $x(t)$ is given by

$$\langle x^2(t,T) \rangle = \int_{t}^{t+T} \langle D(x(s), s) \rangle ds \tag{3.155}$$

If the average is conditional then (3.154) is to be calculated using the transition density p_2. Otherwise, f_2 is required.

The general form of an Ito process is defined by the stochastic integral equation

$$x(t+T) = x(t) + \int_{t}^{t+T} R((s), s) ds + \int_{t}^{t+T} b(x(s), s) dB(s) \tag{3.156}$$

and consists of an arbitrary Martingale plus a drift term. If the transition density has memory of finitely many states in the past, then by definition of the drift and diffusion coefficients that memory should appear explicitly in the coefficients (McCauley, 2008b).

3.9 Ito processes

Locally the sde generates the Ito process. For questions of existence, continuity, and stability of processes, the reader is referred to L. Arnold's informative text (1992) where Lipshitz conditions and Picard's iterative method for stochastic integral equations are discussed along with stability conditions. Existence requires a Lipshitz condition on drift and diffusion combined, and continuity requires that the drift and diffusion coefficients do not grow faster than quadratically in x. Stability means asymptotically stable, so any process with variance unbounded as a function of t is unstable. Stationary processes are stable, for example. The idea of stability is limitation on the size of allowed fluctuations as time increases. The Wiener process is unstable in the sense that the particle is not limited to a finite region of the x-axis as t increases, but escapes to infinity in infinite time. Stability generally requires putting the particle in a box of finite size.

Excepting the most trivial cases, one cannot try realistically to solve for $x(t)$ via Picard's method. The best way to think of solving an Ito process is to find a way to calculate the two-point transition probability density. Before deriving the pdes that generate the transition density, we first derive some other important relations.

Let $R = 0$. The conditional mean square fluctuation is given as

$$\langle x^2(t,T) \rangle_{\text{cond}} = \int_{t}^{t+T} ds \int_{-\infty}^{\infty} dy \, p_2(y,s|x,t) D(y,s) = \int_{t}^{t+T} ds \langle D(x,s) \rangle_{\text{cond}} \quad (3.157)$$

This is corrected by easily derived extra terms when $R \neq 0$, but those terms are of $O(T^2)$ for small T. The point is that we obtain

$$\langle (x(t+T) - x(t))^2 \rangle_{\text{cond}} \approx D(x,t)T \quad (3.158)$$

as T goes to zero. Using the definition

$$\langle (x(t+T) - x(t))^2 \rangle_{\text{cond}} = \int dx dy (x-y)^2 p_2(x, t+T|y, t) \quad (3.159)$$

we obtain

$$D(x,t) \approx \frac{1}{T} \langle (x(t+T) - x(t))^2 \rangle_{\text{cond}} = \frac{1}{T} \int (x - x_0)^2 p_2(x, t+T|x_0, t) dx_0 \quad (3.160)$$

as T vanishes. Likewise,

$$R(x,t) \approx \frac{1}{T} \langle x(t+T) - x(t) \rangle_{\text{cond}} = \frac{1}{T} \int (x - x_0) p_2(x, t+T|x_0, t) dx_0 \quad (3.161)$$

as T vanishes. We've therefore shown that Ito calculus reproduces the standard definitions of drift and diffusion as conditional averages.

3.9.2 Ito's lemma revisited

Ito's lemma for the transformation of general stochastic process $x(t)$ is based on Ito's theorem. Because $(dB)^2 = dt$ we obtain $(dx)^2 = D(x,t)dt$, so that in any twice-differentiable and invertible transformation of variables $y = G(x,t)$ we obtain

$$dy = \frac{\partial G}{\partial t}dt + \frac{\partial G}{\partial x}dx + \frac{1}{2}\frac{\partial^2 G}{\partial x^2}(dx)^2 \qquad (3.162)$$

Substituting, using the sde for dx, we get the sde

$$dy = \left(\frac{\partial G}{\partial t} + R(x,t)\frac{\partial G}{\partial x} + \frac{D(x,t)}{2}\frac{\partial^2 G}{\partial x^2}\right)dt + \frac{\partial G}{\partial x}b(x,t)dB \qquad (3.163)$$

The diffusion coefficient for the process $y(t)$ is therefore $E(y,t) = (G'(G^{-1}(y,t), t))^2 D(G^{-1}(y,t),t)$. For this, the at least twice-differentiable transformation $y = G(x,t)$ must be invertible. The stochastic integral equation for y follows easily as well. Processes where the transformation exists globally and is invertible are topologically equivalent to the original process x. For example, with price p and log return $x = \ln p$, price and returns behavior are topologically equivalent.

Note next that Martingales

$$M(t) = M(t_0) + \int_{t_0}^{t} (\partial G(x(s),s)/\partial x)b(x(s),s)dB(s) \qquad (3.164)$$

can be constructed by requiring that

$$\frac{\partial G}{\partial t} + R(x,t)\frac{\partial G}{\partial x} + \frac{D(x,t)}{2}\frac{\partial^2 G}{\partial x^2} = 0 \qquad (3.165)$$

subject to specific initial or boundary conditions, and that infinitely many different Martingales correspond to the different possible choices of initial or boundary conditions. Equation (3.165) is a backward-time-parabolic pde, a backward-time-diffusion equation because $b^2(x,t) = D(x,t) > 0$.

Next, we derive the forward-time-diffusive pde satisfied by the transition density of an Ito process.

3.9.3 The Fokker–Planck pde

We arrive now at a most important and extremely useful juncture, one basic in statistical physics but seldom mentioned in economics and finance (the reverse is true of the Ito sde). However, our derivation is far more general than the usual textbook one, and makes no assumption that the underlying process is necessarily Markovian.

Consider a twice-differentiable dynamical variable $A(x)$ whose average is finite. The time evolution of A is given by Ito's lemma:

$$dA = \left(R\frac{\partial A}{\partial x} + \frac{D}{2}\frac{\partial^2 A}{\partial x^2}\right)dt + b\frac{\partial A}{\partial x}dB \qquad (3.166)$$

We can calculate the average of A conditioned on $x(t_0) = x_0$ at time t_0 in $x(t) = x_0 + \int R(x,s)ds + \int b(x,s)dB(s)$, if we know the transition density,

$$\langle A(x) \rangle_{\text{cond}} = \int p_2(x,t|x_0,t_0)A(x)dx \qquad (3.167)$$

From

$$\frac{d\langle A(x) \rangle_{\text{cond}}}{dt} = \int \frac{\partial p_2(x,t|x_0,t_0)}{\partial t}A(x)dx \qquad (3.168)$$

and using

$$\langle dA \rangle_{\text{cond}} = \left(\left\langle R\frac{\partial A}{\partial x}\right\rangle_{\text{cond}} + \left\langle \frac{D}{2}\frac{\partial^2 A}{\partial x^2}\right\rangle_{\text{cond}}\right)dt \qquad (3.169)$$

with $d\langle A\rangle/dt$ defined by (3.168), we obtain from (3.169), after integrating twice by parts and assuming that the boundary terms vanish, that

$$\int dx A(x) \left[\frac{\partial p_2}{\partial t} + \frac{\partial(Rp_2)}{\partial x} - \frac{1}{2}\frac{\partial^2(Dp_2)}{\partial x^2}\right] = 0 \qquad (3.170)$$

For arbitrary $A(x)$ this yields

$$\frac{\partial p_2}{\partial t} = -\frac{\partial(Rp_2)}{\partial x} + \frac{1}{2}\frac{\partial^2(Dp_2)}{\partial x^2}. \qquad (3.171)$$

This is the Fokker–Planck pde, or Kolmogorov's second pde (K2), and the Green function is the transition density. Given the transition density and the one-point density we can construct a two-point density $f_2(x,t;y,s) = p_2(x,t|y,s)f_1(y,s)$ where by integration the one-point density satisfies

$$f_1(x,t) = \int f_2(x,t;y,s)dy = \int p_2(x,t|y,s)f_1(y,s)dy \qquad (3.172)$$

and so satisfies the same pde as does p_2, but with an arbitrary choice of initial condition $f_1(x,t_0)$. The transition density satisfies $p_2(x,t_0|x_0,t_0) = \delta(x-x_0)$ and reflects the underlying dynamics while f_1 does not. No Markovian assumption was made, and (3.171) is not an approximation: the Fokker–Planck pde is demanded by the Ito sde (3.166). Note that the line of reasoning here is not the standard one found in other texts (excepting Friedman (1975), in part); our treatment here follows Schulten (1999).

In particular, no assumption was made that R, D, and hence p_2, are independent of memory of an initial state, or of finitely many earlier states. If there is memory, e.g. if $p_1(x,t_0) = u(x)$ and if $D = D(x,t;x_0,t_0)$ depends on one initial state $x_0 = \int xu(x)dx$, then due to memory in p,

$$p_2(x_3,t_3|x_2,t_2) = \frac{\int p_3(x_3,t_3|x_2,t_2;x_1,t_1)p_2(x_2,t_2|x_1,t_1)p_1(x_1,t_1)dx_1}{\int p_2(x_2,t_2|x_1,t_1)p_1(x_1,t_1)dx_1} \quad (3.173)$$

Then by the two-point transition density we must understand that $p_2(x,t|y,s) = p_3(x,t|y,s;x_0,t_1)$. That is, in the simplest case p_3 is required to describe the stochastic process. Memory appears in (3.173) if, for example, at time t_0, $f_1(x,0) = \delta(x-x_0)$ with $x_0 \neq 0$.

It's now quite easy to prove that the sde

$$dx = b(t)dB \quad (3.174)$$

is simply a change of time variable on the Wiener process. Consider scaling solutions of the corresponding pde

$$\frac{\partial f}{\partial t} = \frac{D(t)}{2}\frac{\partial^2 f}{\partial x^2} \quad (3.175)$$

$f(x,t) = \sigma^{-1}(t)F(u), u = x/\sigma(t)$, so that

$$\langle x^2(t) \rangle = \sigma^2(t)\int duu^2 F(u) = \sigma^2(t) \quad (3.176)$$

and $F(u) = (2\pi)^{-1/2}\exp(-u^2/2)$. Scaling requires that

$$D(t) = \frac{d}{dt}\sigma^2(t) \quad (3.177)$$

so that with $dt = D(t)dt$,

$$\tau = \int_0^t D(t)dt = \sigma^2(t) \quad (3.178)$$

we obtain $f(x,t) = g(x,\tau)$ satisfying

$$\frac{\partial g}{\partial \tau} = \frac{1}{2}\frac{\partial^2 g}{\partial x^2} \qquad (3.179)$$

with corresponding sde $dx = dB(\tau)$. The Green function of this pde is the transition density of the Wiener process. Time-translational invariance means that $D(x,t)$ depends on x alone; space-translational invariance means that $D(x,t)$ depends on t alone. The only time- *and* space-translationally invariant Martingale (requiring that $D(x,t) = $ constant) is therefore the Wiener process.

3.9.4 Calculating averages and correlations

We can use Ito's lemma for the calculation of averages. Consider first the average of any dynamical variable $A(x)$ that doesn't depend explicitly on t (e.g. $A(x) = x^2$). From Ito's lemma we obtain

$$dA = (RA'(x) + \frac{D}{2}A''(x))dt + A'bdB \qquad (3.180)$$

Whether the average is conditioned or not we obtain

$$\frac{d\langle A \rangle}{dt} = \langle RA'(x) \rangle + \frac{1}{2}\langle DA''(x) \rangle \qquad (3.181)$$

For example, if $A = x^n$ (and not worrying here about unfinite moments due to fat tails) then

$$\frac{d\langle x^n \rangle}{dt} = n\langle Rx^{n-1} \rangle + \frac{n(n-1)}{2}\langle Dx^{n-2} \rangle \qquad (3.182)$$

Correlations are equally easy to calculate. Suppose one wants to know $\langle A(x(t+T))B(x(t)) \rangle$. For simplicity consider the case where $A(x(0)) = 0$, $B(x(0)) = 0$ and assume that $R = 0$, that $x(t)$ is a Martingale (the reader can easily generalize the result). Then from Ito's lemma

$$A(x(t)) = \int_0^t A''(x(s))D(x(s),s)ds + A''b \bullet \Delta B \qquad (3.183)$$

and the analogous equation for B, we obtain the correlation function for unequal times t and $t + T$. For example, let $A = B = x^2$. Then

$$\langle x^2(t+T)x^2(t)\rangle = \int_0^{t+T}\int_0^t dsdq\langle D(x(s),s)D(x(q),q)\rangle$$
$$+ \left\langle \int_0^t x^4(s)D(x(s),s)\,ds\right\rangle \tag{3.184}$$

The second term on the right-hand side is uninteresting if t is taken to be fixed and T is varied. The quantity (3.184) is used as a "volatility measure" in finance. In the literature it's often taken for granted that memory is necessary to understand the variation of a volatility measure with time lag T, but even Markov models may produce interesting volatility. The reader should by now have become aware of the power of Ito calculus, but we will provide more examples below.

Here's another way to formulate the same result. Using $A(x(t+T))=x^2(t+T)$ in (3.183), multiplying by $x^2(t)$ and averaging we obtain

$$\langle x^2(t+T)x^2(t)\rangle = \langle x^4(t)\rangle + \int_t^{t+T}\langle x^2(t)D(x(s),s)\rangle ds \tag{3.185}$$

3.9.5 Stationary processes revisited

A process is called strongly stationary if densities f_n and transition densities p_n of all orders n are time-translationally invariant and normalizable:

$$f_n(x_1,t_1+T;\ldots;x_n,t_n+T) = f_n(x_1,t_1;\ldots;x_n,t_n) \tag{3.186}$$

For $n=2$ this requires

$$p_2(x_n,t_n|x_{n-1},t_{n-1}) = p_2(x_n,t_n-t_{n-1}|x_{n-1},0) \tag{3.187}$$

and for $n=1$ we must have $f_1(x,t+T)=f_1(x,t)$ independent of t. We can easily produce examples where the stationary one-point density is not normalizable even if (3.187) holds. Such processes are nonstationary.

A *stationary process* demands a normalizable time-translationally invariant one-point density $f_1(x)$, so that the mean $\langle x(t)\rangle$, variance $\sigma^2 = \langle x^2(t)\rangle - \langle x(t)\rangle^2$, and all higher moments are constants, independent of t. The normalizable one-point density describes fluctuations about statistical equilibrium (or a driven steady state), where the equilibrium values of the process are the averages calculated using that density. In equilibrium nothing changes with time.

For a time-translationally invariant process $x(t)$ with normalizable density $f_1(x)$, $p_2(y,t+T|x,t) = p_2(y,T|x,0)$ yields pair correlations

$$\langle x(t+T)x(t)\rangle = \langle x(T)x(0)\rangle \tag{3.188}$$

depending on T alone, independent of t. With $T = 0$ we get $\sigma^2 = \langle x^2(t) \rangle = \langle x^2(0) \rangle =$ constant if $x(0) = 0$. Stationary pair correlations generally exhibit the asymptotic approach to equilibrium. These pair correlations do *not* follow for a time-translational invariant Markov process if the stationary density $f_1(x)$ is not normalizable (an example is provided below) and contradicts the Martingale pair correlations, which depend on t alone independently of the time lag T.

For a time-translationally invariant Martingale

$$x(t+T) = x(t) + \int_t^{t+T} b(x(s))dB(s) \tag{3.189}$$

we obtain

$$\sigma^2 = \langle x^2(t) \rangle = \int_0^t ds \int_{-\infty}^{\infty} dy D(y) p_2(y, s|0, 0) \tag{3.190}$$

which depends on t and generally does not approach a constant for $t \gg 1$ excepting one special case. A Martingale would agree with the stationarity requirement if $\langle x(T)x(0) \rangle = \langle x^2(0) \rangle$, but only the single case of translationally invariant quadratic diffusion yields that result. Generally, a Martingale process is nonstationary.

In contrast with Martingales, the increment autocorrelations of a stationary process do not vanish,

$$\langle x(t, T)x(t, -T) \rangle = \langle x(2T)x(0) \rangle - \sigma^2 \tag{3.191}$$

except in the iid case.

Time-translationally invariant Martingales are defined by the class where $D(x,t) = D(x) \neq$ constant, but where there is no normalizable t-independent one-point density. A normalizable one-point density is provided by $f_1(x, t) = p_2(x, t|0, 0)$. The simplest example is provided by the drift-free lognormal process $dx = xdB(t)$. In Chapter 4, we show that the variance of this process is nonlinear in time t and increases without bound: the process is nonstationary.

More generally, stationary processes may be Markovian, but time-translationally invariant Markov processes are generally not stationary. In the case of an infinite or semi-infinite interval ($b = \infty$) a time-translationally invariant Markov process is generally not stationary because the stationary one-point density is not normalizable, and this is the rule, not the exception. Such a process does not describe fluctuations about statistical equilibrium. In this case a time-dependent mean and the moments are calculated from

$f_1(x,t) = p_2(x,t|0,0)$ with initial condition $f_1(x,0) = \delta(x)$. Here's the simplest example, with $-\infty < x < \infty$.

Consider the Wiener process $dx = dB$ on the entire interval $-\infty < x < \infty$,

$$\frac{\partial g}{\partial t} = \frac{1}{2}\frac{\partial^2 g}{\partial x^2} \tag{3.192}$$

The transition density/Green function is given by the Green function

$$g(x,t|x_0,t_0) = \frac{1}{\sqrt{4\pi\Delta t}} e^{-\frac{(x-x_0)^2}{2\sigma\Delta t}} \tag{3.193}$$

The equilibrium one-point density satisfies $\partial^2 f_1/\partial x^2 = 0$ and is not normalizable, therefore the Wiener process is nonstationary. Another way to see that the process is nonstationary is that $\sigma^2(t) = t$ is never constant. We can take as the normalizable nonequilibrium one-point density $f_1(x,t) = g(x,t|0,0)$.

The lognormal model $dp = \mu p dt + \sigma_1 p dB$ can be solved by transforming to the Gaussian model. With $x = \ln p$, Ito's lemma plus a trivial stochastic integration yields $x = (\mu - \sigma_1^2/2)t + \sigma_1 B$ so that $p(t) = p(0)[\exp(\mu - \sigma_1^2/2)t]\exp\sigma_1 B(t)$. Evaluating $\langle e^{\sigma_1 B}\rangle$ by completing the squares in the Gaussian exponent for the Wiener process yields $\langle p^n(t)\rangle = p^n(0)e^{n[\mu+\sigma_1^2(n-1)/2]t}$. Since the variance grows unbounded as t increases, the process is strongly nonstationary, reflecting the stochastic analog of an unstable dynamical system (Arnold, 1992).

In stark contrast, an example of a process with an asymptotic approach to statistical equilibrium, an asymptotically stationary process, is provided by the OU process:

$$dv = -\beta v dt + \sigma_1 dB \tag{3.194}$$

($\beta > 0$).

In Ito calculus the derivative of a stochastic variable and an ordinary function obeys the ordinary chain rule (proof left to reader) so that

$$e^{-\beta t}d(ve^{\beta t}) = \sigma_1 dB$$

$$v = v_0 e^{-\beta t} + \sigma_1 e^{-\beta t}\int_0^t e^{\beta s}dB(s) \tag{3.195}$$

$$\langle v(t)\rangle = v_0 e^{-\beta t} \to 0, t \to \infty$$

$$\sigma^2 = \langle v(t)^2\rangle - \langle v(t)\rangle^2 = \frac{\sigma_1^2}{2\beta}(1 - e^{-2\beta t}) \to \frac{\sigma_1^2}{2\beta}, t \to \infty \tag{3.196}$$

Stationary processes obey a "fluctuation-dissipation theorem" whereby the friction constant β is determined by the equilibrium fluctuations, yielding a

3.9 Ito processes

relationship between β and σ_1. In equilibrium statistical physics v is the speed of a Brownian particle; the equilibrium density $f(v)$ satisfies

$$\frac{\partial f}{\partial t} = \beta \frac{\partial}{\partial v}(vf) + \frac{\sigma_1^2}{2}\frac{\partial^2 f}{\partial v^2} = 0 \tag{3.197}$$

and is the Maxwell–Boltzmann density (a Gaussian)

$$f(v) \propto e^{-v^2/2kT} \tag{3.198}$$

The fluctuation-dissipation theorem (first noted by Einstein in 1905) yields $\sigma_1^2/\beta = kT$ where k is Boltzmann's constant and T is the absolute/Kelvin temperature of the heat bath, the fluid molecules that the Brownian particle continually collides with. See Kubo *et al.* (1978) on the fluctuation-dissipation theorem.

We can derive the transition density of the OU process by using Ito calculus combined with our knowledge of Gaussian processes. Note that in the solution

$$v(t)e^{\beta t} - v(0) = \sigma_1 \int_0^t e^{\beta q} dB(q) \tag{3.199}$$

the noise term

$$M(t) = \sigma_1 \int_0^t e^{\beta q} dB(q) \tag{3.200}$$

is a Martingale with nonstationary increments because the variance

$$\langle M^2(t) \rangle = \frac{\sigma_1^2}{2\beta}(e^{2\beta t} - 1) \tag{3.201}$$

is nonlinear in t. On the other hand this Martingale is Gaussian because the diffusion coefficient depends on t alone. We can easily construct the transition density. According to (3.119), we have

$$\langle (x - my)^2 \rangle_{\text{cond}} = K(t, s) \tag{3.202}$$

From the standard form

$$v(t) = v(s)e^{-\beta(t-s)} + \sigma_1 e^{-\beta t}\int_s^t e^{\beta q} dB(q) \tag{3.203}$$

we see that $m(t,s) = e^{-(t-s)}$ and so

$$K(t,s) = e^{-2\beta t}\langle M^2(s,t-s)\rangle = \frac{\sigma_1^2}{2\beta}(1-e^{-2(t-s)}) \quad (3.204)$$

where the Martingale increment in (3.200) is given by

$$M(s,t-s) = \sigma_1 \int_s^t e^{2\beta q} dB(q) \quad (3.205)$$

This yields the time-translationally invariant transition density

$$p_2(v,t-s|w,0) = \frac{1}{\sqrt{2\pi\sigma_1^2(1-e^{-(t-s)})/2\beta}} e^{-(v-we^{-\beta(t-s)})/2\sigma_1^2(1-e^{-(t-s)})/2\beta} \quad (3.206)$$

and the pair correlations are given by $\langle x(T)x(0)\rangle = \sigma^2 e^{-\beta T}$.

The transition density is normally derived by solving the Fokker–Planck pde for the OU process. The OU process is asymptotically stationary: a normalizable one-point density $f_1(v) = p_2(v,\infty|0,0)$ representing statistical equilibrium occurs in the limit where $t - s \to \infty$. The reader may check to see that the functions

$$m(t,s) = e^{-(t-s)}$$
$$K(t,s) = \frac{\sigma_1^2}{2\beta}(1-e^{-(t-s)}) \quad (3.207)$$
$$g(t,s) = 0$$

obey the conditions (3.122) for a Gaussian transition density to satisfy the CK equation. We also see that the OU process is Markovian, because there's no memory in either the drift or diffusion term in the sde for the OU process.

3.9.6 Stationary one-point densities

A time-independent one-point density can be derived from the Fokker–Planck pde whenever the drift and diffusion coefficients are t-independent but to qualify as defining a stationary process, the time-independent one-point density must be normalizable. The Fokker–Planck equation expresses local conservation of probability

$$\frac{\partial f}{\partial t} = -\frac{\partial j}{\partial x} \quad (3.208)$$

where the probability current density is

3.9 Ito processes

$$j(x,t) = Rf(x,t) - \frac{1}{2}\frac{\partial}{\partial x}(Df(x,t)) \tag{3.209}$$

Global probability conservation

$$\int_{-\infty}^{\infty} f(x,t)dx = 1 \tag{3.210}$$

requires

$$\frac{d}{dt}\int fdx = \int \frac{\partial f}{\partial t}dx = -j\Big|_{-\infty}^{\infty} = 0 \tag{3.211}$$

Equilibrium solutions may exist only if both $R(x)$ and $D(x)$ are time-independent, and then must satisfy

$$j(x,t) = Rf(x,t) - \frac{1}{2}\frac{\partial}{\partial x}(Df(x,t)) = 0 \tag{3.212}$$

and are given by

$$f(x) = \frac{C}{D(x)}e^{2\int \frac{R(x)}{D(x)}dx} \tag{3.213}$$

with C a constant. The general stationary state, in contrast, follows from integrating (again, only if R and D are t-independent) the first order equation

$$j = R(x)f(x) - \frac{1}{2}\frac{\partial}{\partial x}(D(x)f(x)) = J = \text{constant} \neq 0 \tag{3.214}$$

and is given by

$$f(x) = \frac{C}{D(x)}e^{2\int \frac{R(x)}{D(x)}dx} + \frac{J}{D(x)}e^{2\int \frac{R(x)}{D(x)}dx}\int e^{-2\int \frac{R(x)}{D(x)}dx}dx \tag{3.215}$$

Stationary solutions reflect either statistical equilibrium ($J = \text{constant} = 0$) or the driven time-independent steady state ($j(x) = J = \text{constant} \neq 0$). In the OU process $J = 0$.

Note that time-translationally invariant Martingales generate stationary solutions of the form $f_1(x) = C/D(x)$. These solutions generally are not normalizable on $-\infty < x < \infty$ if the quadratic growth limitation on $D(x)$ for continuity of the stochastic process is met. A stationarity process is not guaranteed for arbitrary time-translationally invariant drift and diffusion coefficients, and this was not pointed out by Kubo, who assumed without discussion that there would be an approach to statistical equilibrium for arbitrary $D(x)$ if $R(x) < 0$.

3.9.7 Stationary increment Martingales

Consider the increment density (3.107) of a time-translationally invariant process. With both p_2 and f_1 independent of t we obtain an increment density $f(z,0,T)$ independent of t. Hence, the increments of stationary processes are also stationary. This does not hold for a general Markov process where p_2 is time-translationally invariant but there is no normalizable time-translationally invariant one-point density. Nor does it generally hold for a Martingale.

What class of Martingales has stationary increments? An arbitrary Martingale increment has the form

$$x(t,T) = \int_t^{t+T} b(x(s),s)dB(s) \qquad (3.216)$$

By time-translational invariance of the Wiener process, we obtain

$$x(t,T) = \int_0^T b(x(s+t),s+t)dB(s) \qquad (3.217)$$

To go further we would need to assume that $x(t+T) = x(t)$ in distribution, i.e. we would have to assume stationary increments under the stochastic integral sign. Actually, we know already that time-translationally invariant Martingales (Martingales with $b(x,t)$ depending on x alone) have nonstationary increments because the variance is not linear in t. The lognormal process $dp = \mu p dt + \sigma_1 p dB(t)$ provides the simplest example, as was shown by the moment derivation above. Since we can rule out the case $D(x,t) = D(x)$ as a candidate for diffusion with stationary increments, what's left?

Let's return to the increment density

$$f(z,t,t+T) = \int p_2(x+z,t+T|x,t)f_1(x,t)dx \qquad (3.218)$$

If we assume time-translational invariance of the transition density,

$$f(z,t,t+T) = \int p_2(x+z,T|x,0)f_1(x,t)dx \qquad (3.219)$$

then this is clearly not enough. To obtain stationary increments we would need to assume in addition that the transition density is space-translationally invariant as well, so that $f(z,t,t+T) = p_2(z,T|0,0)$. But this reduces us to the Wiener process.

Aside from the Wiener process, we do not know how to construct a Martingale model with stationary increments (we ignore the Levy processes here since they aren't diffusive). We've shown above that there is a class of stationary increment processes with long time increment correlations. In Chapter 6 we will produce a Gaussian example from that class, and will see that there *both* p_2 *and* f_1 break time-translational invariance. For nonstationary processes, time-translational invariance is generally not a property of stationary increment processes.

3.10 Martingales and backward-time diffusion

We turn now to a topic seldom treated in the physics and financial math literature. The result is central to the theory of stochastic processes, it goes hand-in-hand with the Fokker–Planck pde and provides the clearest basis for understanding the Black–Scholes model of risk neutral option pricing in Chapter 8.

3.10.1 Kolmogorov's backward-time pde

Consider a diffusive process described by an Ito sde

$$dx = R(x,t)dt + b(x,t)dB(t) \tag{3.220}$$

Consider a twice-differentiable dynamical variable $A(x,t)$. The sde for A is

$$dA = \left(\frac{\partial A}{\partial t} + R\frac{\partial A}{\partial x} + \frac{D}{2}\frac{\partial^2 A}{\partial x^2}\right)dt + b\frac{\partial A}{\partial x}dB \tag{3.221}$$

so that

$$A(x(t+T), t+T) = A(x(t), t) + \int_t^{t+T} \left(\frac{\partial A(x(s), s)}{\partial t} + R\frac{\partial A}{\partial x} + \frac{D}{2}\frac{\partial^2 A}{\partial x^2}\right)ds + \int_t^{t+T} b(x(s), s)\frac{\partial A(x(s), s)}{\partial x}dB(s) \tag{3.222}$$

A Martingale is defined by the conditional average $\langle A \rangle_{cond} = A(x,t)$ where a backward-in-time average is indicated. The backward-time pde, Kolmogorov's first equation, follows directly from requiring that the drift term vanishes,

$$\frac{\partial A(x(s), s)}{\partial t} + R\frac{\partial A}{\partial x} + \frac{D}{2}\frac{\partial^2 A}{\partial x^2} = 0 \tag{3.223}$$

yielding a Martingale

$$A(x(t+T), t+T) = A(x(t), t) + \int_t^{t+T} b(x(s), s) \frac{\partial A(x(s), s)}{\partial x} dB(s) \qquad (3.224)$$

We've made no assumption that A is positive. That is, A is generally not a one-point probability density, $A(x,t)$ is simply any Martingale, and an infinity of Martingales can be so constructed depending on the choice of forward-time initial conditions specified on A (an initial value or boundary value problem backward in time is solved). Let p^+ denote the backward time-transition density of the process (3.223). Because of linearity of the solution of the initial value problem,

$$A(x, t) = \int dy\, p^+(x, t|y, t_0) A(y, t_0) \qquad (3.225)$$

where $A(x,t_0)$ is the forward-time initial condition to be specified. The required transition density therefore satisfies the same pde as the Martingale,

$$0 = \frac{\partial p^+(x, t|y, s)}{\partial t} + R(x, t) \frac{\partial p^+(x, t|y, s)}{\partial x} + \frac{D(x, t)}{2} \frac{\partial^2 p^+(x, t|y, s)}{\partial x^2} \qquad (3.226)$$

where $p^+(x,t|y,t) = \delta(x - y)$. The conditions under which p^+ exists, is unique, and is nonnegative definite are stated in Friedman (1975). Equation (3.226) is called Kolmogorov's first pde (K1). Kolmogorov's backward-time pde is fundamental for understanding Martingale option pricing.

3.10.2 The adjoint pde

Return now to the calculation of averages of dynamical variables via the transition density p_2. Now, for the case where $A(x(t))$ *is* a Martingale (requiring that the drift term in dA vanishes), (3.167) must yield

$$\langle A \rangle_t = \int p(x, t|x_0, t_0) A(x) dx = A(x_0) \qquad (3.227)$$

and since (3.227) cannot differ from (3.225) if the theory is to make any sense, then there must be a connection between the backward and forward time-transition densities p^+ and p_2. Comparing (3.225) with (3.227) we see that p^+ and p_2 must be adjoints,

$$p^+(x, t|y, s) = p_2(y, s|x, t) \qquad (3.228)$$

Consequently, to calculate Martingale option pricing we will need only to solve a certain Fokker–Planck pde.

3.10 Martingales and backward-time diffusion

We've proven elsewhere that the Green functions of K1 and K2 plus boundary conditions imply the Chapman-Kolmogorov equation even if finite memory is present. That the Chapman-Kolmogorov equation makes sense in the face of finite memory follows from

$$p_{k-1}(x_k, t_k | x_{k-2}, t_{k-2}; \ldots; x_1, t_1) = \int dx_{k-1} p_k(x_k, t_k | x_{k-1}, t_{k-1}; \ldots; x_1, t_1)$$
$$p_{k-1}(x_{k-1}, t_{k-1} | x_{k-2}, t_{k-2}; \ldots; x_1, t_1) \quad (3.229)$$

If $p_k = p_n$ for all $k \geq n$, then from (3.229) we obtain the Chapman-Kolmogorov equation in the form

$$p_n(x_n, t_n | x_{n-1}, t_{n-1}; \ldots; x_1, t_1) = \int dy\, p_n(x_n, t_n | y, s; x_{n-2}, t_{n-2}; \ldots; x_1, t_1)$$
$$p_n(y, s | x_{n-1}, t_{n-1}; \ldots; x_1, t_1) \quad (3.230)$$

Explicit examples are shown in McCauley (2008b).

4
Introduction to financial economics

We will begin with several standard ideas from finance: no-arbitrage, the time value of money, and the Modigliani–Miller theorem, and then will introduce the newer concepts of liquidity and reversible trading, market instability, value in uncertain markets, and Black's idea of noise traders. More fundamentally, we'll formulate the efficient market hypothesis (EMH) as a Martingale condition, reflecting a hard-to-beat market. We'll also formulate hypothetical stationary markets, and show that stationary markets and efficient markets are mutually exclusive. A dynamic generalization of the neoclassical notion of value to real, nonstationary markets is presented. We will rely heavily on our knowledge of stochastic processes presented in Chapter 3. Orientation in finance market history is provided in the books by Bernstein (1992), Lewis (1989), Dunbar (2000), and Eichengreen (1996).

4.1 What does no-arbitrage mean?

The basic idea of horse trading is to buy a nag cheap and unload it on someone else for a profit. An analog of horse trading occurs in financial markets, where it's called "arbitrage." The French word sounds more respectable than the Germanic phrase, especially to academics and bankers.[1]

The idea of arbitrage is simple. If gold sells for $1001 in Dubai and for $989 in New York, then traders should tend to sell gold short in Dubai and simultaneously buy it in New York, assuming that transaction costs and taxes are less than the total gain (taxes and transaction costs are ignored to zeroth order in theoretical finance arguments). This brings us to two points. Trading is often performed by using some fraction of borrowed money. In the

---

[1] For a lively description of the bond market in the time of the early days of derivatives, deregulation, and computerization on Wall Street, see *Liar's Poker* by the ex-bond salesman Lewis (1989).

stock market, that's called margin trading and may be highly leveraged. Short selling means borrowing someone else's shares for sale (via a brokerage house, for example) for a fee, and then replacing the shares on or before the expiration date for the sell order.

A basic assumption in standard finance theory (Bodie and Merton, 1998) is that arbitrage opportunities are expected to disappear very quickly because there are many "hungry" and competent professional traders (the civilized analog of hunters) looking systematically for profits. Such traders are normally assumed to be posed to take advantage of any opportunity that presents itself (there is an unstated assumption that the market is liquid enough that fast trading is possible). This leads to the so-called no-arbitrage argument, the so-called "law of one price" (Bodie and Merton, 1998). The idea is that arbitrage occurs on a very short time scale, and on longer time scales equivalent assets will then tend to have more or less the same ask price (or bid price) in different markets (assuming markets with similar tax structure, transaction costs, etc).

Arbitrage arguments are also applied to entirely different assets, like Motorola and Intel. Deciding what is an equivalent asset here is like comparing apples with oranges and can be dangerous because there is no falsifiable basis for equivalence. One must also be careful not to confuse a no-arbitrage condition with the entirely different condition of market equilibrium (market clearing). For example, in Nakamura (2000) the no-arbitrage condition is assumed to represent Adam Smith's Invisible Hand. But consider two geographically separated markets with different prices for the same asset, say Intel. Via arbitrage the price may be lowered in one market and raised in the other, but even if the prices are the same in both markets a positive excess demand will cause the price to increase as time goes on. Therefore, the absence of arbitrage opportunities does not imply either equilibrium or stability. By Adam Smith's Invisible Hand we mean market stability, or an asymptotically stationary market. The Invisible Hand is a synonym for an unspecified mechanism that moves prices toward a statistical equilibrium distribution.

Arbitrage, carefully and precisely stated, is based on the comparison of spatially separated prices of *the same asset at a single time t*. To make clear that "no-arbitrage" is not an equilibrium condition (has nothing to do with time-translational invariance), we identify the correct analogy with an underlying symmetry and invariance principle. To see that a no-arbitrage condition doesn't imply vanishing total excess demand for an asset, consider two spatially separated markets with two different price distributions for the same asset. If enough traders go long in one market and short in the other, then the market price distributions may be brought into agreement. Even then, if there is positive excess demand for the asset then the average price of the asset will continue increasing with time, so that there is no equilibrium. So, markets that are far from equilibrium can satisfy the no-arbitrage condition. Here, econophysicists have the advantage over economists: we understand the difference between time-translational invariance on the one hand, and spatial translational/rotational invariance on the other. In particular, continuous symmetries are intimately connected with conservation laws (Wigner, 1967).

In order to understand the geometric meaning of the no-arbitrage condition, consider a spatial distribution of markets with different price distributions at each location, i.e. the stock AMD has different prices in New York, Tokyo, Johannesburg, Frankfurt, London, and Moscow. That is, the price distribution $g(p,X,t)$ depends on price p, market location X, and time t. It is now easy to formulate the no-arbitrage condition in the language of statistical physics. The no-arbitrage condition means spatial homogeneity and isotropy of the price distribution (to within transaction, shipping and customs fees, and taxes). In the ridiculously oversimplified case of a uniform market distribution over the globe, "no-arbitrage" would be analogous to rotational invariance of the price distribution, and to two-dimensional translational invariance locally in any tangent plane (from Boston to New York, for example). The price distribution is *not* required to be stationary, so market equilibrium/market clearing is not achieved merely by the lack of arbitrage opportunities. Given this, how can we define the underlying "value" of an asset?

The terms "overpriced" and "underpriced" are often heard and read in the financial news. But to determine whether an asset is overpriced or underpriced we would need a notion of "value." Do paper assets like money or stocks really admit an observable intrinsic or fundamental value, or any useful notion of value other than the current market price?

4.2 Nonfalsifiable notions of value

A debt of one Dollar owed today is worth less to the lender if payment is deferred for a year. If the annual bank interest rate is r, then one Dollar promised to be paid now but paid instead after a year is worth only $\$1/(1+r)$ to the recipient today, or $PV = FV/(1+r)$ where $p(t_0) = PV$ is present value and $p(t) = FV$ is future value. In finance texts this is called "the time value of money." For n discrete time intervals Δt with interest rate r for each interval, we have $p(t) = p(t_0)(1 + r\Delta t)^n$. This is also called "discounting." In continuous time $p(t_n) = p(t_0)e^{r\Delta t}$ so that present value is $p(t_0) = p(t)e^{-r\Delta t}$.

The time value of money is determined by the ratio of two prices at two different times. Consider either money or any other asset, like a stock or a house. Is there an underlying "true value" of the asset, a fundamental price at a single time t? The answer is *"Jain"* (German for "yes-no"). First, even in standard economics thinking the value of an asset is not a uniquely defined idea: there are at least five different definitions of "value" in finance theory. The first is book value. The second uses the replacement price of a firm (less taxes owed, debt, and other transaction costs). These first two definitions are respected by market fundamentalists, and can apparently be useful, judging from Warren Buffett's successes. But it's rare that market prices for companies with good future prospects fall to book value, although the writer knew one such case in 1974 and by 1975 the price had quadrupled (anti-pollution laws had just been passed, and the company had just begun business in that area). Instead, we will concentrate on the standard ideas of value from finance theory, ideas not used by successful traders. A still popular idea among some theorists is the old idea of dividends and returns discounted infinitely into the future for a financial asset like a stock or bond. This reminds us vaguely of the neo-classical condition of "infinite foresight" on the part of agents. The fourth idea of valuation, due to Modigliani and Miller, is somewhat more realistic and is discussed in part 4.4 below.

The idea of dividends and returns discounted infinitely into the future is not falsifiable because it makes impossible demands on human knowledge. Here's the formal definition:

Starting with the total return given by the gain $R\Delta t$ due to price increase with no dividend paid in a time interval Δt, and using the small returns approximation, we have

$$\Delta x = \ln p(t)/p(t_0) \approx \Delta p/p \qquad (4.1)$$

or

$$p(t + \Delta t) \approx p(t)(1 + R\Delta t) \qquad (4.2)$$

But paying a dividend d at the end of a quarter ($\Delta t =$ one quarter) reduces the stock price, so that for the nth quarter

$$p_n = p_{n-1}(1 + R_n) - d_n \qquad (4.3)$$

If we solve this by iteration for the implied fair value of the stock at time t_0 then we obtain

$$p(t_0) = \sum_{k=1}^{\infty} \frac{d_n}{1 + R_n} \qquad (4.4)$$

whose convergence assumes that p_n goes to zero as n goes to infinity. This reflects the assumption that the stock is only worth its dividends, an assumption of little or no practical use in investing, especially as Modigliani and Miller have explained that dividends don't matter in the valuation of a firm. Robert Shiller (1999) uses this formal definition of value in his theoretical discussion of the market efficiency in the context of rational vs irrational behavior of agents, but we will not follow that discussion in this book. What is ignored in the above model and in neo-classical economics is precisely what we focus on here, that markets are based on uncertainty about the future. Nothing could be a bigger waste of time for an investor than trying to guess a flow of future dividends.

Finally, in the neo-classical model where money/liquidity is excluded because uncertainty was systematically and unrealistically deleted, "value" is the price-label at which equilibrium occurs. There, 100% of all agents agree on value. Undervalued and overvalued would be well defined were money allowed, but money is not allowed and cannot be introduced. The generalization of the neo-classical model to an uncertain market is a stationary market. We will show that "value" can be identified in a stationary market, and that the notion can be extended to nonstationary markets as well. The latter is the idea of value that we will deal with in this text.

4.3 The Gambler's Ruin

The Gambler's Ruin is a useful idea, as it provides advice about making many small bets in the market compared with a single large bet.

Consider any game with two players (you and the stock market, for example). Let d denote a gambler's stake, and D the house's stake. If borrowing is not possible then $d + D = C =$ constant is the total amount of capital. Let R_d denote the probability that the gambler goes broke, in other words the probability that $d = 0$ so that $D = C$. Assume a fair game; for example, each player bets on the outcome of the toss of a fair coin. Then

$$R_d = \frac{1}{2}R_{d+1} + \frac{1}{2}R_{d-1} \tag{4.5}$$

with boundary conditions $R_0 = 1$ (ruin is certain) and $R_C = 0$ (ruin is impossible). To solve (4.5), assume that R_d is linear in d. The solution is

$$R_d = \frac{D}{C} = 1 - \frac{d}{C} \tag{4.6}$$

Note first that the expected gain for either player is zero,

$$\langle G \rangle = -dR_d + D(1 - R_d) = 0 \tag{4.7}$$

representing a fair game on the average: for many identical repetitions of the same game, the net expected gain for either the player or the bank vanishes, meaning that sometimes the bank must also go broke in a hypothetically unlimited number of repetitions of the game. In other words, in infinitely many repeated games the idea of a fair game would re-emerge: neither the bank nor the opponent would lose money on balance. However, in finitely many games the house, or bank, with much greater capital has the advantage; the player with much less capital is much more likely to go broke. Therefore if you play a fair game many times and start with capital $d < D$ you should expect to lose to the bank, or to the market, because in this case $R_d > 1/2$. An interesting side lesson taught by this example that we do not discuss here is that, with limited capital, if you "must" make a gain "or else," then it's better to place a single bet of all your capital on one game, even though the odds are that you will lose. By placing a single large bet instead of many small bets you improve your odds (Billingsley, 1983).

But what does a brokerage house have to do with a casino? The answer is: quite a lot. Actually, a brokerage house can be understood as a full-service casino (Lewis, 1989; Millman, 1995). Not only will they place your bets. They'll lend you the money to bet with, on margin, up to 50%. However, there is an important distinction between gambling in a casino and gambling in a financial market. In the former, the probabilities are fixed: no matter how many people bet on red, if the roulette wheel turns up black they all lose. In the market, the probability that you win increases with the number of people making the same bet as you. If you buy a stock and many other people buy the same stock afterward then the price is driven upward. You win if you sell before the others get out of the market. That is, in order to win you must (as Keynes pointed out) guess correctly what other people are going to do before they do it. This would require having better than average information about the economic prospects of a particular business, and also the health of the economic sector as a whole. Successful traders like Soros and Buffett are examples of agents with much better than average knowledge. They don't defeat the EMH, they go around it.

4.4 The Modigliani–Miller argument

We define the "capital structure" of a publicly held company as the division of financial obligations into stocks and bonds. The estimated value[2] of a firm

---

[2] One might compare this with the idea of "loan value," the value estimated by a bank for the purpose of lending money.

is given by $p = B + S$ where B is the total debt and S is the equity, also called market capitalization. Defined as $B + S$, market value p is measurable because we can find out what is B, and $S = p_s N_s$ where N_s is the number of shares of stock outstanding at price p_s. For shares of a publicly traded firm like intc, one can look up both N_s and p_s on any discount broker's website. The Modigliani–Miller (M & M, meaning Franco Modigliani and Merton Miller) theorem asserts that capital structure doesn't matter, that the firm's market value p (what the firm would presumably sell for on the open market, were it for sale) is independent of the ratio B/S. Liquidity of the market is taken for granted in this discussion (otherwise there may be no buyers, in which case the M & M price estimate is not useful) in spite of the fact that huge, global companies like Exxon and GMC rarely change hands: the capital required for taking them over is typically too large.

Prior to the M & M (1958) theorem it had been merely assumed without proof that the market value p of a firm must depend on the fraction of a firm's debt vs its equity, B/S. In contrast with that viewpoint, the M & M theorem seems intuitively correct if we apply it to the special case of buying a house or car: how much one would have to pay for either today is roughly independent of how much one pays initially as downpayment (this is analogous to S) and how much one borrows to finance the rest (which is analogous to B). From this simple perspective, the correctness of the M & M argument seems obvious. Let's now reproduce M & M's "proof" of their famous theorem.

Their "proof" is based on the idea of comparing "cash flows" of equivalent firms. M & M neglect taxes and transaction fees and assume a very liquid market, one where everyone can borrow at the same risk-free interest rate. In order to present their argument we can start with a simple extrapolation of the future based on the local approximation ignoring noise

$$\Delta p \approx r p \Delta t \tag{4.8}$$

where $p(t)$ should be the price of the firm at time t. This equation assumes the usual exponential growth in price for a risk-free asset like a money market account where r is fixed. Take the expected return r to be the market capitalization rate, the expected growth rate in value of the firm via earnings (the cash flow), so that Δp denotes earnings over a time interval Δt. In this picture p represents the value of a firm *today* based on the market's expectations of its *future* earnings $<\Delta p>$ at a later time $t + \Delta t$. To arrive at the M & M argument we concentrate on

$$p \approx \langle \Delta p \rangle / r \tag{4.9}$$

where p is to be understood as today's estimate of the firm's net financial worth based on $\langle \Delta p \rangle = E$ and r is the expected profit and expected rate of increase in value of the firm over one unit of time, one quarter of a year. If we take $\Delta t =$ one quarter in what follows, then E denotes expected quarterly earnings. With these assumptions, the "cash flow" relation $E = pr$ yields that the estimated fair price of the firm today would be

$$p = E/r \qquad (4.10)$$

where r is the expected rate of profit per quarter and E is the expected quarterly earnings. Of course, in reality we have to know E at time $t + \Delta t$ and p at time t and then r can be estimated. Neither E nor r can be known in advance and must either be estimated from historic data (assuming that the future will be like the past) or else guessed on the basis of new information. In the relationship $p = B + S$, in contrast, B and S are always observable at time t. B is the amount of money raised by the firm for its daily operations by issuing bonds and S is the market capitalization, the amount of money raised by issuing shares of stock.

Here comes the main point: M & M want us to assume that estimating E/r at time t is how the market arrives at the observable quantities B and S. To say the least, this is a very questionable proposition. In M & M's way of thinking, if the estimated price E/r differs from the market price $p = B + S$ then there is an arbitrage opportunity. M & M assume that there is no arbitrage possible, so that the estimated price E/r and the known value $B + S$ *must* be the same. Typically of neo-classical economists, M & M mislabel the equality $B + S = E/r$ as "market equilibrium," although the equality has nothing to do with equilibrium, because in equilibrium nothing can change with time.

In setting $B + S = p = E/r$, M & M make an implicit assumption that the market collectively "computes" p by estimating E/r, although E/r cannot be known in advance. That is, an implicit, unstated model of agents' collective behavior is assumed without empirical evidence. The assumption is characteristic of neo-classical thinking.[3] One could try to assert that the distribution of prices, which is in reality mainly noise (and is completely neglected in M & M), reflects all agents' attempts to compute E/r, but it is doubtful that this is what agents really do, or that the noise can be interpreted as any definite form of computation. In reality, agents do not seem to behave like ideally rational bookkeepers who succeed in obtaining all available information in numerical bits. Instead of bookkeepers and calculators, one can more accurately speak

---

[3] The market would have to behave trivially like a primitive computer that does only simple arithmetic, and that with data that are not known in advance. Contrast this with the complexity of intellectual processes described in Hadamard (1945).

of agents, who speculate about many factors like the "mood" of the market, the general economic climate of the day triggered by the latest news on unemployment figures, etc., and about how other agents will interpret that data. One also should not undervalue personal reasons like financial constraints, or any irrational look into the crystal ball. The entire problem of agents' psychology and herd behavior is swept under the rug with the simple assumptions made by M & M, or by assuming optimizing behavior. Of course, speculation is a form of gambling: in speculating one places a bet that the future will develop in a certain way and not in alternative ways. Strategies can be used in casino gambling as well, as in blackjack and poker. In the book *The Predictors* (Bass, 1999) we learn how the use of a small computer hidden in the shoe and operated with the foot leads to strategies in roulette as well.

This aside was necessary because when we can agree that agents behave less like rational computers and more like gamblers, then M & M have ignored something important: the risk factor, and risk requires the inclusion of noise[4] as well as possible changes in the "risk-free" interest rate which are not perfectly predictable and are subject to political tactics by the Federal Reserve Bank.

Next, we follow M & M to show that dividend policy should not affect net shareholders' wealth in a perfect market, where there are no taxes and transaction fees. The market price of a share of stock is just $p_s = S/N_s$. Actually, it is p_s and N_s that are observable and S that must be calculated from this equation. Whether or not the firm pays dividends to shareholders is irrelevant: paying dividends would reduce S, thereby reducing p_s to $p'_s = (S-\delta S)/N_s$. This is no different in effect than paying interest due quarterly on a bond. Paying a dividend is equivalent to paying no dividend but instead diluting the market by issuing more shares to the same shareholders (the firm could pay dividends in shares), so that $p'_s = S/(N_s + \delta N_s) = (S-\delta S)/N_s$. In either case, or with no dividends at all, the net wealth of shareholders is the same: dividend policy affects share price but not shareholders' wealth. Note that we do not get $p_s = 0$ if we set dividends equal to zero, in contrast with (4.4).

Here's a difficulty with the picture we've just presented: although the M & M argument assumes perfect liquidity, liquidity in reality has been ignored (because liquidity is noise). Suppose that the market for firms is not liquid, because most firms are not traded often or in volume. Also, the idea of characterizing a firm or asset by a single price doesn't make sense in practice unless bid/ask spreads are small compared with both bid and ask prices.

---

[4] Ignoring noise is the same as ignoring risk, the risk is in price fluctuations. Also, as F. Black pointed out "noise traders" provide liquidity in the market.

Estimating fair price p independently of the market in order to compare with the market price $B + S$ and find arbitrage opportunities is not as simple as it may seem (see Bose (1999) for an application of equation (4.10) to try to determine if stocks and bonds are mispriced relative to each other). In order to do arbitrage you would have to have an independent way of making a reliable estimate of future earnings E based also on an assumption of what is the rate r during the next quarter. Then, even if you use this guesswork to calculate a "fair price" that differs from the present market price and place your bet on it by buying a put or call, there is no guarantee that the market will eventually go along with your sentiment within your prescribed time frame. For example, if you determine that a stock is overpriced then you can buy a put, but if the stock continues to climb in price then you'll have to meet the margin calls, so the Gambler's Ruin may break your bank account before the stock price falls enough to exercise the put. This is qualitatively what happened to the hedge fund Long Term Capital Management (LTCM), whose collapse in 1998 was a danger to the global financial system (Dunbar, 2000). Remember, there are no springs in the market, only unbounded diffusion of stock prices with nothing to pull them back to your notion of "fair value."

To summarize, the M & M argument that $p = B + S$ is independent of B/S makes sense in some cases,[5] but the assumption that most agents uniformly can compute what they can't know, namely E/r to determine a single fair price p, does not hold water. The impossibility of using then-existing finance theory to make falsifiable predictions led Black via the Capital Asset Pricing Model (CAPM) to discover a falsifiable model of options pricing, which (as he pointed out) can be used to value corporate liabilities. We will present the CAPM in the next chapter. CAPM was essentially the earliest falsifiable contribution to finance market theory.

We next turn to what M & M neglected: the noise that represents the liquidity in the market.

4.5 Excess demand in uncertain markets

We begin by asserting, in agreement with the idea of prices determined by supply and demand, that

$$\frac{dp}{dt} = \varepsilon(p, t) \qquad (4.11)$$

---

[5] For a very nice example of how a too small ratio S/B *can* matter, see pp. 188–190 in Dunbar (2000). Also, the entire subject of Value at Risk (VaR) is about maintaining a high enough ratio of equity to debt to stay out of trouble while trading.

where ε is the excess demand at time t. Market clearing means that $dp/dt = 0$, requiring an equilibrium market. We will show that the assumptions found in the literature of market clearing in finance markets are unfounded and wrong, even in a hypothetical stationary market as we point out below.

An uncertain market is described as a stochastic price process, and if we assume a market composed of drift plus a Martingale then excess demand takes on the form

$$dp = r(p,t)dt + \sqrt{D(p,t)}dB \qquad (4.12)$$

We begin here with the first finance market model by Osborne (1958).

In a noise-free market like a bank deposit we have $r(p,t)=\mu p$ where μ is the bank's interest rate. To generate Osborne's 1958 observation that stock prices were approximately lognormally distributed, we need $D(p,t) = \sigma_1^2 p^2$,

$$dp = \mu p dt + \sigma_1 p dB \qquad (4.13)$$

This was used by Black, Scholes, and Merton to price options falsifiably in 1973, and forms the basis for "financial engineering" today. The model was not falsified until after the 1987 stock market crash. Here, average market clearing $\langle dp/dt \rangle = 0$ is impossible. *Exact* market clearing is always impossible under noise (uncertainty), but *average* market clearing would be possible if and only if the market were a stationary one. We'll show this in the section below on stationary markets.

By Ito's lemma the returns $x = \ln p$ are Gaussian distributed in the lognormal model,

$$dx = (\mu - \sigma_1^2/s)dt + \sigma_1^2 dB \qquad (4.14)$$

The earliest model of a real finance market is nonstationary; it describes a far-from-equilibrium market. To verify this, notice that the variance and all moments of p increase exponentially with time. From the returns

$$x(t) = (\mu - \sigma_1^2/s)t + \sigma_1^2 B(t) \qquad (4.15)$$

and using $p(t) = p(0)e^x$ to obtain the solution

$$p(t) = p(0)e^{(\mu - \sigma_1^2/2)t + \sigma_1 B(t)} \qquad (4.16)$$

the moments are easily calculated by using the one-point Gaussian distribution of B,

$$\langle p^n(t) \rangle = p^n(0)e^{n(\mu - \sigma_1^2(n-1)/2)t} \qquad (4.17)$$

The excess demand vanishes on the average but fluctuates; the model has no approach to stationarity. The measure of the fluctuations

$\sigma^2(t) = \langle p^2(t) \rangle - \langle p(t) \rangle^2$ increases exponentially with time. The lognormal pricing model does not fit the data quantitatively, but is partly qualitatively correct, in the sense that real finance markets are nonstationary. We'll learn that finance markets demand a non-Gaussian nonstationary model

$$dp = rp dt + \sqrt{p^2 d(p,t)} dB \qquad (4.18)$$

one where $d(p,t)$ is both strongly nonlinear and nonseparable in p and t. We'll see in Chapter 6 that this viewpoint is useful for modeling real finance markets.

4.6 Misidentification of equilibrium in economics and finance

There are at least five wrong definitions of equilibrium in the economics and finance literature. The first three definitions would be correct were markets stationary, but economic processes are known to be nonstationary: there is (1) the idea of equilibrium fluctuations about a drift in price, requiring a stationary noise source. Then (2) there is the related notion that market averages describe equilibrium quantities (Fama, 1970). Assumption (3) is widespread in the literature, and is the notion that the CAPM describes "equilibrium" prices (Sharpe, 1964). Again, this definition fails because (as we'll see in the next chapter) the parameters in the CAPM vary with time because finance markets are nonstationary. (4) Black (1989) claimed that "equilibrium dynamics" is described by the Black–Scholes equation. That was equivalent to assuming that (i) normal liquid market returns are Gaussian-Markov distributed, and (ii) that "no-arbitrage" is the same idea as market equilibrium. Absence of arbitrage opportunities is also (mis)identified as "equilibrium" in Bodie and Merton (1998). Finally, there is the idea (5) that the market and stochastic models of the market define sequences of "temporary price equilibria" (Föllmer, 1995). We now proceed to deconstruct definition (5).

The clearest discussion of "temporary price equilibria" is provided by Föllmer (1995). In this picture excess demand can vanish but prices are still fluctuating. Föllmer expresses the notion by trying to define an "equilibrium" price for a sequence of time intervals (very short investment/speculation periods Δt), but the price so defined is not constant in time and is therefore not an equilibrium price. He begins by stating that an equilibrium price would be defined by vanishing total excess demand, $\varepsilon(p) = 0$. He then claims that the condition defines a sequence of "temporary price equilibria," even though the time scale for a "shock" from one "equilibrium" to another would be on the order of a second: the "shock" is nothing but the change in price due to the execution of a new buy or sell order. Föllmer's choice of language sets the stage for encouraging the reader to believe that market prices are, by definition, "equilibrium" prices. In line with this expectation, he next invents a hypothetical excess demand for agent i over time interval $[t, t+\Delta t]$ that is logarithmic in the price,

$$\varepsilon_i(p) = \alpha_i \ln(p_i(t)/p(t)) + \Delta x_i(t, \Delta t) \tag{4.19}$$

where $p_i(t)$ is the price that agent i would be willing to pay for the asset during speculation period Δt. The factor $x_i(t,\Delta t)$ is a "liquidity demand": agent i will not buy the stock unless he already sees a certain amount of demand for the stock in the market. This is a nice idea: the agent looks at the number of limit orders that are the same as his and requires that there should be a certain minimum number before he also places a limit order. By setting the so-defined total excess demand $\varepsilon(p)$ (obtained by summing (4.18) over all agents) equal to zero, one obtains the corresponding equilibrium price of the asset

$$\ln p(t) = \left(\sum_i (\alpha_i \ln p_i(t) + \Delta x_i(t))\right) / \sum_i \alpha_i \tag{4.20}$$

In the model p_i is chosen as follows: the traders have no sense where the market is going so they simply take as their "reference price" $p_i(t)$ the last price demanded in (4.19) at time $t - \Delta t$,

$$p_i(t) = p(t - \Delta t) \tag{4.21}$$

This yields

$$\begin{aligned}\ln p(t) &= \left(\sum_i (\alpha_i \ln p(t - \Delta t) + \Delta x_i(t, \Delta t))\right) / \sum_i \alpha_i \\ &= \ln p(t - \Delta t) + \Delta x(t, \Delta t)\end{aligned} \tag{4.22}$$

If we assume next that the liquidity demand $\Delta x(t,\Delta t)$, which equals the log of the "equilibrium" price increments, executes Brownian motion then we obtain a contradiction: the excess demand (4.20), which is logarithmic in the price p and was assumed to vanish does not agree with the total excess demand defined by the right-hand side of (4.18), which does not vanish, because with $\Delta x = (R - \sigma^2/2)\Delta t + \sigma \Delta B$ we have $dp/dt = r + \sigma dB/dt = \varepsilon(p) \neq 0$. The price $p(t)$ so defined is not an equilibrium price because the resulting lognormal price distribution depends on the time.

A related misuse of the word "equilibrium" appears in Muth's original definition of rational expectations in Chapter 10. Muth wrote demand/consumption and supply (using our notation) as

$$D(\delta p, t) = \beta \delta p(t)$$
$$S(\delta p, t) = \gamma \langle \delta p \rangle_{\text{subj}} + u(t) \tag{4.23}$$

where δp is supposed to be the deviaton from an equilibrium price, $\langle \delta p \rangle_{\text{subj}}$ is the subjectively expected price deviation at time t, and $\varepsilon(t)$ is noise representing the uncertainty of the agents. The idea is to try to understand how producers' expectations correspond to future prices. Setting $D = S$ Muth obtains

$$\delta p(t) = \frac{\gamma}{\beta} \langle \delta p \rangle_{\text{subj}} + \frac{1}{\beta} u(t) \tag{4.24}$$

The result is self-contradictory: from an equilibrium assumption $D = S$ we derive a time-varying price δp. Physics referees would have balked at such an outlandish claim. In physics we have ideas like "local thermodynamics equilibrium" where the temperature and other variables evolve slowly, but not suddenly on a short time scale.

4.7 Searching for Adam Smith's Unreliable Hand

The idea of Adam Smith's Invisible Hand is that markets should tend toward equilibrium, requiring that market equilibrium must (a) exist, and (b) be stable. This requires that the total excess demand for an asset vanishes on the average and that the average asset price and variance are constants.

The OU model

$$dp = -|\mu|p dt + \sigma_1 dB \tag{4.25}$$

with negative interest rate $\mu < 0$ would provide us with a simple model of Adam Smith's stabilizing Invisible Hand. Statistical equilibrium is achieved as t increases. Unfortunately, there is no evidence for such behavior. Asset markets are described qualitatively correctly by the lognormal pricing model

$$dp = \mu p dt + \sigma_1 p dB \tag{4.26}$$

where, even if the interest rate is negative, the model is nonstationary. Here, the variable diffusion coefficient wins over the restoring force and destabilizes the motion.

The Fokker–Planck equation for the lognormal model

$$\frac{\partial g}{\partial t} = -\mu \frac{\partial}{\partial p}(pg) + \frac{\sigma_1^2}{2}\frac{\partial^2}{\partial p^2}(p^2 g) \qquad (4.27)$$

has the time-invariant solution

$$g(p) = C/p^{1+2\mu/\sigma_1^2} \qquad (4.28)$$

which is not normalizable over $0 < p < \infty$, therefore statistical equilibrium does not exist for this model. As we showed in Chapter 3, the variance grows exponentially with t, for example, reflecting loss of knowledge about the prices as t increases. The lognormal process is a time-translationally invariant nonstationary Markov process. Stated otherwise, the Gibbs entropy of this process $S = -\int g \ln g \, dp$ increases without bound.

Statistical equilibrium can be achieved in this model by imposing price controls $p_1 \leq p \leq p_2$. Mathematically, this is represented by reflecting walls at the two end points (one can set $p_1 = 0$ but $p_2 < \infty$ is required), the problem of a particle in a box. In that case, the most general solution of the Fokker–Planck equation is given by the equilibrium solution plus terms that die exponentially as t goes to infinity (Stratonovich, 1963). The spectrum of the Fokker–Planck operator that generates the eigenfunctions has a discrete spectrum for a particle in a box, and the lowest eigenvalue vanishes. It is the vanishing of the lowest eigenvalue that yields equilibrium asymptotically. When the prices are unbounded, the lowest eigenvalue still vanishes but the spectrum is continuous, and equilibrium does not follow. The main point is that the mere mathematical *existence* of a statistical equilibrium solution of the Fokker–Planck equation does not guarantee that time-dependent solutions of that equation will converge to that statistical equilibrium as time goes to infinity unless the stationary solution is normalizable. In this example, Adam Smith's hands are not invisible, but have the form of stiff barriers that limit prices.

We show in the next section that the detrended nonstationary lognormal process, which is nonstationary on the interval $0 \leq p \leq \infty$, describes a hypothetical efficient market. We then show in the section afterward that a stationary model like the price-controlled process above violates the conditions for market efficiency.

4.8 Martingale markets (efficient markets)

In discussing the EMH, we restrict our modeling to a *normal liquid market*. The EMH describes a market that is either very hard or perhaps impossible to beat (McCauley et al., 2007a) and is inapplicable to a market crash. Since we would have to exploit correlations in order to beat a market, the EMH means

4.8 Martingale markets (efficient markets)

that there are no *easy-to-find* correlations or patterns that can be exploited systematically for profit. A Markovian Martingale market would be unbeatable in this sense. Real liquid markets may not be Markovian Martingales; they may be merely very hard but not impossible to beat.

First, we want to deduce the possible form of an efficient market process from the condition that the past provides no knowledge of the future at the level of pair correlations. Higher-order correlations are then left unspecified. In all that follows, we assume that detrending is possible and that the time series under consideration have been detrended. Given our discussion of detrending in Chapter 3, we can hardly avoid deducing a Martingale process.

To formulate the dynamics of hard-to-beat markets we assume that the increment autocorrelations vanish, where by increments we mean $x(t,T) = x(t + T) - x(t)$, $x(t,-T) = x(t) - x(t - T)$. The statement that trading during an earlier time interval provides no signals for traders in a later nonoverlapping time interval *at the level of pair correlations* is simply

$$\langle (x(t_1) - x(t_1 - T_1))(x(t_2 + T_2) - x(t_2)) \rangle = 0 \tag{4.29}$$

If there is no time interval overlap, $[t_1 - T_1, t_1] \cap [t_2, t_2 + T_2] = \emptyset$, where $\emptyset$ denotes the empty set on the line. This is a much less restrictive condition than assuming that the increments are statistically independent on the one hand, or that the detrended market returns are Markovian on the other. The condition (4.29) is necessary but insufficient for a drift-free Markov process. This insufficiency permits market memory at a level beyond pair correlations, in principle, but the necessity makes the market look like a Markovian Martingale at the level of pair correlations or simple averages.

Consider any stochastic process $x(t)$ where the increments are uncorrelated, where (4.29) holds. From this condition we obtain the autocorrelation function for positions (returns). Let $t > s$, then

$$\langle x(t)x(s) \rangle = \langle (x(t) - x(s))x(s) \rangle + \langle x^2(s) \rangle = \langle x^2(s) \rangle > 0 \tag{4.30}$$

since $x(s) - x(t_0) = x(s)$, so that $\langle x(t+T)x(t) \rangle = \langle x^2(t) \rangle$ is simply the variance in x at the earlier time t. *This condition is equivalent to a Martingale process:*

$$\int dy y p_2(y, t + T | x, t) = x \tag{4.31}$$

$$\langle x(t + T)x(t) \rangle = \int\int dx dy xy p_2(y, t + T | x, t) f_1(x, t)$$
$$= \int x f_1(x, t) dx \left(\int y dy p_2(y, t + T | x, t) \right) = \int x^2 f_1(x, t) dx \tag{4.32}$$

Mandelbrot (1966) originally proposed the Martingale as a model for the EMH on the basis of simple averages in x, but we have deduced the Martingale property from a two-point condition, the lack of increment autocorrelations. Note also that (4.30) can be interpreted as asserting that earlier returns have no correlation with future gains.

Next, we discover an extremely important point for data analysis and modeling. Combining

$$\left\langle (x(t+T) - x(t))^2 \right\rangle = \langle x^2(t+T) \rangle + \langle x^2(t) \rangle - 2\langle x(t+T)x(t) \rangle \quad (4.33)$$

with (4.32) we get

$$\left\langle (x(t+T) - x(t))^2 \right\rangle = \langle x^2(t+T) \rangle - \langle x^2(t) \rangle \quad (4.34)$$

which depends on both t and T, excepting the rare case where $\langle x^2(t) \rangle$ is linear in t. *Uncorrelated increments are generally nonstationary.* Notice further that (4.34) states that

$$\sigma^2(t+T) = \langle x^2(t,T) \rangle + \sigma^2(t) \quad (4.35)$$

That is, $\sigma^2(t+T) > \sigma^2(t)$, the variance increases with time, statistical equilibrium cannot be approached unless $\sigma(t)$ approaches a constant limit. Since a Martingale has the form

$$x(t) = x(0) + \int_0^t \sqrt{D(x(s),s)} dB(s) \quad (4.36)$$

the variance taken about $x(0)$ is

$$\sigma^2(t) = \int_0^t \langle D(x(s),s) \rangle ds \quad (4.37)$$

and does not approach a constant as t increases. A Martingale is a nonstationary stochastic process. We'll show in the next section that the pair correlations of an efficient market (a hard-to-beat market) conflict with those of hypothetical stationary markets. An efficient market is nonstationary, is far from equilibrium. This has not been understood in financial economics: either real markets provide falsifiable evidence for an approach to statistical equilibrium at long time, or else they do not. So far, no such evidence has been produced.

The Martingale interpretation of the EMH is interesting because technical traders assume that certain price sequences give signals either to sell or buy. In principle, that is permitted in a Martingale. A particular price sequence

4.8 Martingale markets (efficient markets)

$(p(t_n), \ldots, p(t_1))$, were it quasi-systematically to repeat, can be encoded as returns $(x_n, \ldots, x_1)$ so that a conditional probability density $p_n(x_n|x_{n-1}, \ldots, x_1)$ could be interpreted as providing a risk measure to buy or sell. By "quasi-repetition" of the sequence we mean that $p_n(x_n|x_{n-1}, \ldots, x_1)$ is significantly greater than the corresponding Markovian prediction. Typically, technical traders make the mistake of trying to interpret random price sequences quasi-deterministically, which differs from our interpretation of "technical trading" based on conditional probabilities (see Lo *et al.* (2000) for a discussion of technical trading claims, but based on a non-Martingale, nonempirically based model of prices). With only a conditional probability for "signaling" a specific price sequence, an agent with a large debt-to-equity ratio can easily suffer the Gambler's Ruin. In any case, we can offer no advice about technical trading because the existence of market memory has not been established (the question is left open by the analysis of Lo *et al.*). Liquid finance markets are effectively Markovian Martingales after ten minutes of trading (Chapter 7). We next review the idea of the EMH as it appears typically in economics discussions.

The strict interpretation of the EMH is that there are no correlations, no patterns of any kind, that can be employed *systematically* to beat the average return $\langle R \rangle$ reflecting the market itself: if one wants a higher return, then one must take on more risk (in the French–Fama way of thinking, "omniscient agents" are assumed who neutralize all information up until time t_1). A Markovian Martingale market is unbeatable, it has no systematically repeated patterns, no memory to exploit. We argue that the stipulation should be added that in discussing the EMH we should consider only normal, liquid markets (a normal liquid market is defined precisely below). Otherwise, "Brownian" market models do not apply to describe the market dynamics. Liquidity, the "money bath" created by the noise traders whose behavior is reflected in the diffusion coefficient, is somewhat qualitatively analogous to the idea of the heat bath in thermodynamics: the second-by-second fluctuations in $x(t)$ are created by the continual "noise trading."

Historically, Mandelbrot had proposed the idea of the EMH as a Martingale condition, but discussed only simple averages, not pair correlations. Fama then took Mandelbrot's proposal seriously and tried to test finance data at the simplest level for a fair game condition. Fama made a mathematical mistake (see the first two of three unnumbered equations at the bottom of p. 391 in Fama, 1970) that has become propagated in the literature. He wrongly concluded in his discussion of Martingales as a fair game condition that $\langle x(t+T)x(t) \rangle = 0$. Here's his argument, rewritten partly in our notation. Let $x(t)$ denote a "fair game." With the initial condition chosen as $x(t_0) = 0$, we have the unconditioned expectation $\langle x(t) \rangle = \int x dx f_1(x,t) = 0$ (there is no drift). Then the so-called "serial covariance" is given by

$$\langle x(t+T)x(t) \rangle = \int x dx < x(t+T) >_{\text{cond}(x)} f_1(x,t). \tag{4.38}$$

Fama states that this autocorrelation vanishes because $\langle x(t+T) \rangle_{\text{cond}} = 0$. This is impossible: by a fair game we mean a Martingale, the conditional expectation is

$$\langle x(t+T) \rangle_{\text{cond}} = \int y dy p_2(y, t+T|x,t) = x = x(t) \neq 0 \tag{4.39}$$

and so Fama should have concluded instead that $\langle x(t+T)x(t) \rangle = \langle x^2(t) \rangle$ as we showed in the last section. Vanishing of (4.38) would be true of statistically independent returns but is violated by a "fair game." Can Fama's argument be saved? Suppose that instead of $x(t)$ we would try to use the *increment* $x(t,T) = x(t+T) - x(t)$ as variable. Then $\langle x(t,T)x(t) \rangle = 0$ for a Martingale. However, Fama's argument still would not be generally correct because $x(t,T)$ cannot be taken as a "fair game" variable unless the variance is linear in t.

In our discussion of the EMH we have not followed the economists' tradition of discussing three separate forms (weak, semi-strong, and strong (Skjeltorp, 2000)) of the EMH, where a nonfalsifiable distinction is made between three separate classes of traders. Normal market statistics overwhelmingly (with high probability) reflect the noise traders (Black, 1986), so we consider *only* normal liquid markets and ask whether noise traders produce signals that one might be able to trade on systematically. The question of whether insiders, or exceptional traders like Buffett and Soros, can beat the market probably cannot be tested scientifically: even if we had statistics on such exceptional traders, those statistics would likely be too sparse to draw a firm conclusion. Furthermore, it is not clear that they beat liquid markets; some degree of illiquidity seems to play a significant role there. Effectively, or with high probability, there is only one type of trader under consideration here, the noise trader. The question that we pose is whether, given a Martingale created by the noise traders, a normal liquid market can be beaten systematically at some higher level of correlation than pair correlations. In a word, Buffett and Soros are not noise traders.

4.9 Stationary markets: value and inefficiency

A neo-classical equilibrium deterministic market is a barter system. Money/liquidity does not exist. "Value" is the "price label" at which goods and services are exchanged. Undervalued and overvalued are well-defined ideas,

4.9 Stationary markets: value and inefficiency

but profit is disallowed by lack of money; exchange of goods and services is allowed only at the equilibrium point.

Real markets are noisy. Uncertainty dominates real life. The relaxation of the neo-classical equilibrium straitjacket to permit the least harmful sort of uncertainty leads to a stationary price process. A stationary process describes fluctuations about statistical equilibrium (or a steady state), in which equilibrium is described by a *normalizable* time-invariant one-point density of returns $f_1(x)$. All simple averages are calculated from the equilibrium density, so that nothing changes with time at the level of simple averages. Any attempted definition of "equilbrium" that contradicts this does not describe statistical equilibrium (equilibrium means time-translational invariance with an invariant, normalizable one-point density). We can now identify "value" in a hypothetical stationary market.

In neo-classical economics, value is the price label where barter occurs, and 100% of all agents agree on value. We can generalize this in a useful way. "Value" in an uncertain (fluctuating, noisy) market is the price assigned by the largest fraction of traders to an asset. This price is consensus value, meaning the most probable value, the price where the returns density $f_1(x,t)$ peaks, and this holds whether a market is stationary or nonstationary. In a stationary market, value so identified is constant, does not change with time. We can refer to this as "value under uncertainty." Noise represents agents' uncertainty, and only a small fraction of traders (those with price expectations near the peak of f_1) agree on "value." In a hypothetical stationary market, "overvalued" and "undervalued" are useful, observable ideas because value stands still and the process is recurrent; what goes up must come down and vice versa. It is exactly time-translational invariance that makes such a market inefficient; stationarity makes the market violate the EMH at the level of pair correlations. All earlier economic theorizing about stationarity has missed this key point (McCauley, 2008a), because economists are aware of market efficiency but are married very unfortunately to the unrealistic notion that markets should clear.

In a stationary process, densities f_n and transition densities p_n of all orders n are time-translationally invariant,

$$f_n(x_1, t_1 + T; \ldots; x_n, t_n + T) = f_n(x_1, t_1; \ldots; x_n, t_n) \qquad (4.40)$$

and $p_2(x_n, t_n | x_{n-1}, t_{n-1}) = p_2(x_n, t_n - t_{n-1} | x_{n-1}, 0)$ as well. A stationary process also requires time-translational invariance of a *normalizable* one-point density $f_1(x)$, so that the mean $\langle x(t) \rangle$, variance $\sigma^2 = \langle x^2(t) \rangle - \langle x(t) \rangle^2$, and all higher moments are constants, independent of t. The one-point density describes fluctuations about statistical equilibrium where the equilibrium values of the process are the averages calculated using that density. In equilibrium nothing changes with time. But there is a subtle point that must be appreciated. Some stationary processes are Markovian (the OU process is an example), but time-translationally invariant Markov processes are generally not stationary.

For a time-translationally invariant Markov process a stationary one-point density f_1 can be derived via the Fokker–Planck pde, but the stationary density generally is not *normalizable* unless the process is confined to a box of finite size, $-\infty < a \leq x \leq b < \infty$. The Wiener and related lognormal processes provide the simplest example. In the case of an infinite or semi-infinite interval ($b = \infty$) a time-translationally invariant Markov process is generally not stationary because the stationary one-point density is not normalizable, and this is the rule, not the exception. Such a process does not describe fluctuations about statistical equilibrium. In this case a time-dependent mean and the moments are calculated from $f_1(x,t) = p_2(x,t|0,0)$ with initial condition $f_1(x,0) = \delta(x)$. Again, the lognormal process is the canonical example of a time-translationally invariant nonstationary Markov process. Next we explain how and why a stationary market would contradict the EMH.

Consider a stationary process $x(t)$. Here, $f_1(x,t) = f_1(x)$ is normalizable and time-translational invariance of the transition density $p_2(y, t+T|x,t) = p_2(y, T|x, 0)$ yields pair correlations

$$\langle x(t+T)x(t) \rangle = \langle x(T)x(0) \rangle \tag{4.41}$$

depending on T alone, independent of t or $\sigma^2 = \langle x^2(t) \rangle = \langle x^2(0) \rangle =$ constant with $x(0) = 0$. This result does *not* follow for a time-translational invariant Markov process where $f_1(x)$ is not normalizable. It contradicts the Martingale condition (4.30), where for a Martingale the pair correlations depend on t alone, independent of the time lag T (a drift-free Markov process is a Martingale, with one singular exception). For a time-translationally invariant Martingale

$$x(t+T) = x(t) + \int_{t}^{t+T} \sqrt{D(x(s))} dB(s) \tag{4.42}$$

we obtain

$$\sigma^2 = \langle x^2(t) \rangle = \int_0^t ds \int_{-\infty}^{\infty} dy D(y) p_2(y,s|0,0) \tag{4.43}$$

which depends unavoidably on t. A Martingale is a nonstationary stochastic process. *This means that an efficient market, a hard-to-beat market, cannot be stationary.*

The increment correlations of a stationary process do not vanish; instead we obtain

$$\langle x(t,T)x(t,-T)\rangle = \langle x(2T)x(0)\rangle - \sigma^2 \qquad (4.44)$$

yielding pair correlations that can in principle be traded on for profit. The increments are stationary; the increment autocorrelations do not decay with time.

Without market stationarity there is no way to identify a time-invariant *numeraire*. In option pricing, the so-called "risk-free asset" is taken as the standard against which other prices are measured, and this is taken to be a currency, but currencies are subject to FX fluctuations and inflation. Since the gold standard was abandoned, there has been no currency stability and no time-invariant standard of "value." On the gold standard, the time-invariant *numeraire* was measured in physical units, grams or ounces. Today, the financial equivalent of one gram of measure does not exist for money.

A fluctuation-dissipation theorem can be developed for a class of stationary processes (Kubo *et al.*, 1978), relating the friction coefficient (the analog or return μ) to equilibrium fluctuations. Stationary processes can be used to describe equilibrium statistical physics, from whose time-invariant averages thermodynamics is derived. Were markets stationary, a thermodynamics of economics might in principle make sense. But real markets are nonstationary, and no meaningful thermodynamics of economics has been constructed (McCauley, 2004). With a stationary process the Gibbs entropy can be constructed from the one-point distribution and properly reflects disorder. Historic attempts in economics to base entropy on utility fail miserably, because disorder/uncertainty is completely barred from neo-classical theory. Entropy can only be based on disorder. The Gibbs entropy of the market distribution can be trivially constructed, but does not stabilize, and economic analogies with other physical quantities like energy and free energy do not exist. In particular, since utility is constructed for systems with perfect order, utility bears no relation to entropy.

4.10 Black's "equilibrium": dreams of recurrence in the market

In the short paper "Noise," Fischer Black (1986) discusses three topics: price, value, and noise.[6] He states that price is random and observable whereas value is random and unobservable. He asserts boldly that, because of noise,

---

[6] We recommend the short paper "Noise" by Fischer Black, who wrote and thought very clearly. He died too early to receive the Nobel Prize along with Myron Scholes and Robert Merton. See especially the entertaining NOVA video *The Trillion Dollar Bet*, www.pbs.org/wgbh/nova/stockmarket/

price deviates from value but always returns to value (he introduced the phrase "noise traders" in this paper). He regards price and value as roughly the same if price is within twice value. There is only one problem: he never defines what he means by "value." Black's considerations would have made sense in part, were the price process stationary. We'll point out in Chapter 7 that even stationary densities generally are beyond the reach of empirical analysis, so would a stationary density exist we might have trouble locating the peak accurately.

Black apparently believed the neo-classical economists' ideas of "equilibrium," which he called "beautiful." He should have realized the conflict with his own nonstationary pricing model, the Black–Scholes model, but apparently did not. We can only guess what Black may have thought, but the following argument would explain Black's claims about price and value. The market, as Osborne taught us, consists of unfilled limit book orders that are step functions. One can see these step functions evolving in time on the website 3DCharts.com, and one can consult Nasdaq Level 2 for detailed numerical information. If we would assume that market equilibria exist and are stable, as neo-classical economics teaches, then every limit book would have a daily clearing price, namely, the equilibrium price, where total supply exactly matches total demand. Were the clearing price to exist, then it could be taken to define "value." Were the equilibrium stable, then price would always tend to return to value no matter how far price would deviate from value. The trades occur in discrete prices, and a discrete stationary process is necessarily recurrent (Kac, 1949, 1959b). Unfortunately, the evidence suggests strongly that markets are far from equilibrium, and are nonstationary.

4.11 Value in real, nonstationary markets

In a stationary market the generalization of perfectly agreed-on "value" (the equilibrium price) can be taken as the location of the peak of the price density. Such a market describes statistical equilibrium, and we now generalize the idea of value to general nonstationary markets. The definition is the same: the peak of the price density locates consensus value.

Suppose that price obeys $dp = \mu p dt + p\sqrt{d(p,t)}dB$. A price series can only be detrended multiplicatively: if $S = pe^{-\mu t}$ then $dS = S\sqrt{e(S,t)}dB$ is a Martingale, where $e(S,t) = d(p,t)$. Let V represent the most probable price. The local approximation about the price V at time t_0 is described by the local Gaussian density

$$g(S,t) \approx (e(V,t_0)T)^{-1/2} e^{-(S-V)^2/2e(V,t_0)T} \qquad (4.45)$$

where $t - t_0 = T$. With T small but finite, $g(V,t) \approx (e(V,t_0)T)^{-1/2}$ is maximum if the most probable price V locates the minimum of the returns diffusion coefficient $D(x - \mu t, t) = e(S,t)$. Or, if there is only one price scale in the process then we can expect that the consensus price $p_c(t) = p_c e^{\mu t}$ locates the minimum of the diffusion coefficient. This is difficult to prove rigorously in general, our argument here is heuristic, but we will give an explicit example in Chapter 7 where it holds rigorously, the case of scaling Martingales.

The existence of a price scale, the consensus price, is essential. In the transformation from price to log returns we must define $x(t) = \ln(p(t)/V(t))$ where $V(t)$ is a price that makes the argument of the logarithm dimensionless. In principle $V(t)$ may be arbitrary, can be taken as constant, but there's then a further unanswered question: what sets the price scale of the log return in the returns diffusion coefficient $D(x,t)$? We will largely consider models where $D(x,t)$ increases with $|x|$, so the question is: for what price is D a minimum? Again, if there is but one price scale, then the consensus price sets the scale for the diffusion coefficient as well. We therefore generally will assume in what follows that $x(t) = \ln(p(t)/p_c(t))$ so that the corresponding increments are $x(t,T) = \ln(e^{-\mu T} p(t+T)/p(t))$. These increments do not represent detrended returns, because the returns sde corresponding to a detrended price S is $dx = -D(x,t)dt/2 + b(x,t)dB$. We would have approximately detrended returns if and only if we could ignore the variable drift term proportional to D in the sde. Detrending a time series is intimately connected with the question of consensus price. We will return to this question in Chapter 7. In any case, in a nonstationary market there is no tendency of price to return to consensus value, there is only the tendency to diffuse away from value. Placing bets in a nonstationary market on the assumption that prices will recur would amount to taking on high risk against the advice of the market.

Fischer Black apparently believed in the neo-classical notion of stable market equilibrium (so did Merton and Scholes, as the history of LTCM (Dunbar, 2000) makes clear): Black argued that prices will always tend to return to "value." Here, he was wrong: *there is no tendency for prices to "return to value."* Because market dynamics are unstable/nonstationary, price always diffuses away from "value," there being no "springs" in the market to pull prices back to value. In contrast, Soros (1998, 2008) is correct: financial markets are dynamically unstable.

4.12 Liquidity, noise traders, crashes, and fat tails

By a normal liquid market for a very frequently traded stock like Intc, Aapl, or Nok we mean a market where the bid/ask spread is tiny compared with the bid and ask prices, or with the price of the last executed trade. The reader can best get a feeling for this by checking bid, ask, and last trade prices on a discount broker site on a computer. An essential condition for the applicability of stochastic processes is that the market consists of an adequate "liquidity bath." A normal liquid market in a frequently traded stock approximates this. By a normal liquid market we mean the following: a money/liquidity bath is assumed, in part analogous to the heat bath in statistical physics, where approximately reversible trades are possible via your discount broker in real time over the shortest time intervals Δt on the order of a few seconds on your Mac or PC. An approximately reversible trade is one where you can reverse your trade in a few seconds with only a small loss or gain. This assumes that the brokerage firm executes limit orders for small trades in real time (1000 shares of the above mentioned stocks is an example of a small trade). This works in detail in the following way. You have Y shares of Intc to sell. You check for the last bid price, and place a limit order to sell Y shares of Intc at a slightly lower price. An illiquid market is one with large bid/ask spreads, like housing, carpets, or cars, where trades occur far less frequently and with much lower volume than in financial markets.

Our definition of liquidity assumes implicitly that the future will obey the same statistics as the past. This ignores the surprises that characterize complexity in simple dynamical systems. We ignore the possibility of a fundamental shift in the market distribution. Option pricing, for example, is based on this assumption.

Fischer Black (1986) has taught us that liquidity is provided by the noise traders. It's the noise traders who make it possible for us to place a buy limit order at a price slightly above the ask price and have the order executed in real time (if your discount broker is a good one). The noise traders are represented by the noise term $\sqrt{D(x,t)}dB(t)$ in the Ito sde. Liquidity can be understood as entropy, the Gibbs entropy of the one-point distribution is $S = -\int f_1 \ln f_1 dx$ and increases with time. The noise traders constitute the market "with measure one": *uncertain of "value," they buy and sell frequently*: a financial market is essentially noise because most traders don't have inside or other useful knowledge to trade on.

Fat tails don't describe market crashes. We'll show in Chapter 6 that fat tails may describe large returns that occur during normal liquid markets. In contrast, a market crash is a *liquidity drought* where the noise traders can't sell because there are effectively no buyers. This is described qualitatively to zeroth order by $R << 0$ with $D(x,t) \approx 0$. The systematic degradation of the Dollar, as this is being written, can likely be understood by a liquid market with systematic drops in R.

When we refer to "market price" we make an implicit assumption of adequate liquidity. A liquid market is one with many rapidly executed trades in both directions, always far from equilibrium, where bid/ask spreads are small compared with price. This allows us to define "market price" in real time as the price at the last trade. Examples of liquid markets are well-traded stocks and bonds, and foreign exchange of currencies like the Euro, Dollar, and Yen, so long as large buy/sell orders are avoided, and so long as there's no market crash.

The former trader George Soros, who bet heavily against the Bank of England and won, asserts that the market is always wrong. He tries to explain what he means by this in his book *The Alchemy of Finance* (1994) and more recent books, but like a baseball batter trying to explain how to hit the ball, Soros was much better at winning than at explaining how he wins. He discusses the difference between science and self-fulfilling expectations, and introduces the interesting idea of a perception gap between what we believe to be market reality, and market reality, which is not knowable in real time. A bubble is an example of a self-fulfilling expectation.

4.13 Long-term capital management

A main theme of this book is that there are no forces to cause a market to tend toward an equilibrium state. Markets are nonstationary, and nonstationary processes are neither ergodic nor recurrent. There is no statistical evidence for Adam Smith's Invisible Hand. Recurrence of a stationary price process would have provided the most general possible description of Adam Smith's Hand but stationarity is not to be found in real markets, only in the markets hypothesized in macroeconomic and econometric texts.

The dramatically failed hedge fund LTCM assumed that deviations from Black–Scholes option pricing would always return to historic market averages (Dunbar, 2000). This was an implicit assumption of ergodicity. Initially, the fund made a lot of money for several years during the mid-nineties by betting on small-fluctuation "mispricing." LTCM had two Nobel Prize-winning neo-classical economists on its staff, Merton and Scholes. They apparently assumed some form of market stability in spite of the fact that the model used by them to price options is nonstationary. Finally, LTCM suffered the Gambler's Ruin during a long time-interval large deviation. For a very interesting story of how, in contrast, a group of physicists who do not believe in equilibrium and stability placed bets in the market during the nineties and are still in business, see *The Predictors* (Bass, 1998).

The hedge fund used the idea of a fair price for options. The Black–Scholes option price can be shown to represent a Martingale. The model therefore provides a basis for arbitrage: if one finds "mispricing" in the form of option prices that violate Black–Scholes, then a bet can be placed that the deviation from the Black–Scholes prediction will disappear, that the market will eliminate these "efficiencies" via arbitrage. That is, Black–Scholes assumes that the market is efficient in the sense of the EMH, which is fine so long as liquidity does not dry up. But LTCM placed bets on deviations from historic behavior that grew in magnitude instead of disappearing over a relatively long time interval precisely because, as Dunbar described, their positions were so large that they literally became the market in certain assets. As the spread widened they continued to place more bets, assuming that returns would spring back to historic values on a relatively short time scale, even though there were no buyers in sight. That's how they suffered the Gambler's Ruin. Both the recurrence property of a stationary process and the price fairness of a Martingale process presume adequate liquidity, otherwise those stochastic models fail.

5
Introduction to portfolio selection theory

5.1 Introduction

Everyone would like to know how to pick winning stocks, but there exists no reliable mathematical theory, nor is a guaranteed qualitative method of success[1] available to us. Given one risky asset, how much should one bet on it? According to the Gambler's Ruin, we should bet the whole amount if winning is essential for survival. If, however, one has a time horizon beyond the immediate present, then maybe the amount gambled should be less than the amount required for survival in the long run. Given two or more risky assets we can ask Harry Markowitz's question: can we choose the fractions invested in each in such a way as to minimize the risk, where risk is defined by the standard deviation of the expected return? This is the beginning of the analysis of the question of risk vs reward via diversification and assumes normal liquid markets.

This chapter is written on the assumption that the future will be statistically like the past, that the historic statistical price distributions of financial markets are adequate to predict future expectations like option prices. This assumption will fail miserably during a liquidity crunch, and also after the occurrence of any surprise that changes market psychology permanently.

5.2 Risk and return

A so-called risk-free asset has been defined historically as one with a fixed interest rate, like a CD, money market account, or treasury bill (this definition is based on the assumption that the currency in question is not

---

[1] According to Warren Buffett, more or less: pick a stock that has good earnings prospects. Don't be afraid to buy when the market is low. Do be afraid to buy when the market is high. This advice goes against that inferred from the EMH. Soros (2008) offers his qualitative idea of how the market works ("reflexivity").

systematically debased, so that at the time of writing the Euro and Yen serve the purpose far better than the Dollar). Barring financial disaster, you're certain to get your money back, plus interest. A risky asset is one that fluctuates in price, one where retrieving the capital cannot be guaranteed, especially over the long run. In all that follows we work with returns $x = \ln(p(t)/p(0))$ instead of prices p.

Averages

$$R = \langle x \rangle = \langle \ln(p(t)/p(0)) \rangle \tag{5.1}$$

are understood always to be taken with respect to the empirical distribution unless we specify that we are calculating for a particular model distribution in order to make a point. The empirical distribution is not an equilibrium one because its moments change with time without approaching any constant limit. Finance texts written from the standpoint of neo-classical economics assume "equilibrium," but statistical equilibrium would require time-independence of the empirical distribution, and this is not found in any financial market. The Gaussian model of returns used by Black and Scholes is an example of a nonequilibrium distribution.

Consider first a single risky asset with expected return R_1 combined with a risk-free one with known return R_0. Let f denote the fraction invested in the risky asset. The fluctuating return of the portfolio is given by $x = fR_1 + (1-f)R_0$ and so the expected return of the portfolio is

$$R = fR_1 + (1-f)R_0 = R_0 + f\Delta R \tag{5.2}$$

where $\Delta R = R_1 - R_0$. The portfolio standard deviation, or root mean square fluctuation (rmsf), is given as

$$\sigma = f\sigma_1 \tag{5.3}$$

where

$$\sigma_1 = \left\langle (x - R_1)^2 \right\rangle^{1/2} \tag{5.4}$$

is the standard deviation of the risky asset. We can therefore write

$$R = R_0 + \frac{\sigma}{\sigma_1} \Delta R \tag{5.5}$$

which we will generalize later to include many uncorrelated and also correlated assets.

In this simplest case the relation between return and risk is linear (Figure 5.1): the return is linear in the portfolio standard deviation. The greater the expected return, the greater the risk. If there is no chance of

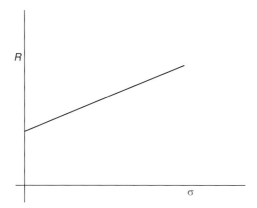

Figure 5.1 Return R vs "risk"/standard deviation σ for a portfolio made up of one risky asset and one risk-free one.

return then a trader or investor will not place the bet corresponding to buying the risky asset.

We've seen in Chapter 4 that the Gambler's Ruin advises us to put all our eggs in one basket if it's a matter of survival. Here, we concentrate on the opposite advice: diversification of bets. In particular, the law of large numbers can be used to show how to reduce risk (measured as variance) in a normal liquid market in a portfolio of n risky assets. The Strategy of Bold Play and the Strategy of Diversification provide mutually exclusive answers to different questions.

5.3 Diversification and correlations

Consider next n uncorrelated assets; the x_k are all assumed hypothetically to be distributed statistically independently. The expected return would be given by

$$R = \sum_{k=1}^{n} f_k R_k \tag{5.6}$$

and the mean square fluctuation by

$$\sigma^2 = \left\langle \left(\sum f_k x_k - R \right)^2 \right\rangle = \sum f_k^2 \sigma_k^2 \tag{5.7}$$

where f_k is the fraction of the total budget that is bet on asset k.

As a special case consider a portfolio constructed by dart throwing (a favorite theme in Malkiel (1996), who assumes statistical independence where it doesn't apply):

$$f_k = 1/n \tag{5.8}$$

Let σ_1 denote the largest of the σ_k. Then

$$\sigma \leq \frac{\sigma_1}{\sqrt{n}} \tag{5.9}$$

This shows how the variance/uncertainty could in principle be reduced by diversification with a statistically independent choice of assets. But statistically independent assets are hard or impossible to find. For example, automobile and auto supply stocks are correlated within the sector, computer chip and networking stocks are correlated with each other, and there are also correlations across different sectors due to general business and political conditions.

Consider a portfolio of two assets with historically expected return given by

$$R = fR_1 + (1-f)R_2 = R_2 + f(R_1 - R_2) \tag{5.10}$$

and risk-squared by

$$\sigma^2 = f^2\sigma_1^2 + (1-f)^2\sigma_2^2 + 2f(1-f)\sigma_{12} \tag{5.11}$$

where

$$\sigma_{12} = \langle (x_1 - R_1)(x_2 - R_2) \rangle \tag{5.12}$$

describes the correlation between the two assets. Eliminating f via

$$f = \frac{R - R_2}{R_1 - R_2} \tag{5.13}$$

and solving

$$\sigma^2 = \left(\frac{R-R_2}{R_1-R_2}\right)^2 \sigma_1^2 + \left(1 - \frac{R-R_2}{R_1-R_2}\right)^2 \sigma_2^2 + 2\frac{R-R_2}{R_1-R_2}\left(1 - \frac{R-R_2}{R_1-R_2}\right)\sigma_{12} \tag{5.14}$$

for reward R as a function of risk σ yields a parabola opening along the σ-axis, which is shown in Figure 5.2.

Now, given any choice for f we can combine the risky portfolio (as fraction w) with a risk-free asset to obtain

$$R_T = (1-w)R_0 + wR = R_0 + w\Delta R \tag{5.15}$$

With $\sigma_T = w\sigma$ we therefore have

$$R_T = R_0 + \frac{\sigma_T}{\sigma}\Delta R \tag{5.16}$$

5.3 Diversification and correlations

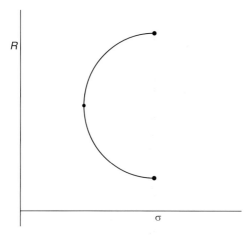

Figure 5.2 The efficient portfolio, showing the minimum risk portfolio as the left-most point on the curve.

The fraction $w = \sigma_T/\sigma$ describes the level of risk that the agent is willing to tolerate. The choice $w = 0$ corresponds to no risk at all, $R_T = R_0$, and $w = 1$ corresponds to maximum risk, $R_T = R_1$.

Next, let's return to equations (5.14)–(5.16). There's a minimum risk portfolio that we can locate by using (5.14) and solving

$$\frac{d\sigma^2}{dR} = 0 \qquad (5.17)$$

Instead, because R is proportional to f, we can solve

$$\frac{d\sigma^2}{df} = 0 \qquad (5.18)$$

to obtain

$$f = \frac{\sigma_2^2 - \sigma_{12}}{\sigma_1^2 + \sigma_2^2 - 2\sigma_{12}} \qquad (5.19)$$

Here, as a simple example to prepare the reader for the more important case, risk is minimized independently of expected return. Next, we derive the so-called "tangency portfolio," also called the "efficient portfolio" (Bodie and Merton, 1998). We can minimize risk with a given expected return as constraint, which is mathematically the same as maximizing the expected return for a given fixed level σ of risk. This leads to the so-called efficient and tangency portfolios. First, we redefine the reference interest rate to be the risk-free rate. The return relative to R_0 is

$$\Delta R = R - R_0 = f_1 \Delta R_1 + f_2 \Delta R_2 \tag{5.20}$$

where $\Delta R_k = R_k - R_0$ and where we've used the constraint $f_1 + f_2 = 1$. The mean square fluctuation of the portfolio is

$$\sigma^2 = \langle \Delta x^2 \rangle = f_1^2 \sigma_1^2 + f_2^2 \sigma_2^2 + 2f_1 f_2 \sigma_{12} \tag{5.21}$$

Keep in mind that the five quantities R_k, σ^2_k, and σ_{12} should in principle be calculated from empirical data and will vary with time. Next, we minimize the mean square fluctuation subject to the constraint that the expected return (5.20) is fixed. In other words we minimize the quantity

$$H = \sigma^2 + \lambda(\Delta R - f_1 \Delta R_1 - f_2 \Delta R_2) \tag{5.22}$$

with respect to the fs, where λ is the Lagrange multiplier. This yields

$$\frac{\partial H}{\partial f_1} = 2f_1 \sigma_1^2 + 2f_2 \sigma_{12} - \lambda \Delta R_1 = 0 \tag{5.23}$$

and likewise for f_2. Using the second equation to eliminate the Lagrange multiplier λ yields

$$\lambda = \frac{2f_2 \sigma_2^2 + 2f_1 \sigma_{12}}{\Delta R_2} \tag{5.24}$$

and so we obtain

$$2f_1 \sigma_1^2 + 2f_2 \sigma_{12} - \frac{\Delta R_1}{\Delta R_2}(2f_2 \sigma_2^2 + 2f_1 \sigma_{12}) = 0 \tag{5.25}$$

Combining this with the second corresponding equation (obtained by permuting indices in (5.25)) we can solve for f_1 and f_2. Using the constraint $f_2 = 1 - f_1$ yields

$$f_1 = \frac{\sigma_2^2 \Delta R_1 - \sigma_{12} \Delta R_2}{(\sigma_1^2 - \sigma_{12})\Delta R_2 + (\sigma_2^2 - \sigma_{12})\Delta R_1} \tag{5.26}$$

and likewise for f_2. This pair (f_1, f_2), so calculated, defines the efficient portfolio of two risky assets. In what follows we denote the expected return and mean square fluctuation of this portfolio by R_e and σ_{ee}.

If we combine the efficient portfolio as fraction w of a total investment including the risk-free asset, then we obtain the so-called tangent portfolio

$$R_T = R_0 + w \Delta R_e \tag{5.27}$$

where $\Delta R_e = R_e - R_0$ and w is the fraction invested in the efficient portfolio, the risky asset. With $\sigma_T = w \sigma_e$ we have

5.4 The CAPM portfolio selection strategy

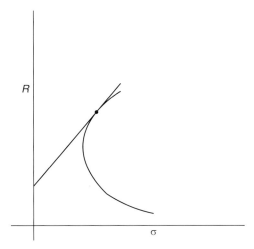

Figure 5.3 The tangency portfolio.

$$R_T = R_0 + \frac{\sigma_T}{\sigma_e}\Delta R_e \qquad (5.28)$$

The result is shown as Figure 5.3. Tobin's separation theorem (Bodie and Merton, 1998), based on the tangency portfolio (another Nobel Prize in economics), corresponds to the trivial fact that nothing determines w other than the agent's psychological risk tolerance, or the investor's preference: the value of w is given by free choice. Clearly, a younger person far from retirement may sensibly choose a much larger value for w than an older person who must live off the investment. Unless, of course, the older person is in dire straits and must act boldly or else face the financial music. But as happened in the late 1990s it can also go otherwise: older people with safe retirement finances gambled by following the fad of momentum trading via home computer. That was in the late days of the dot.com bubble. As one financial advisor recently said about the subprime mortgage fiasco, one knows it's a bubble, but it's hard to quit before the music stops. See also Soros (2008) for the difficulty of shorting the market while the bubble continues to expand.

We turn next to the standard and first model of diversification that takes into account stock correlations in a normal liquid market, CAPM.

5.4 The CAPM portfolio selection strategy

The Capital Asset Pricing Model (CAPM) is very general: it assumes no particular distribution of returns and is consistent with *any* distribution with finite first and second moments. Therefore, in this section, we generally assume the empirical distribution of returns. The CAPM (Varian, 1992) is not, as is often claimed (Sharpe, 1964), an equilibrium model because the distribution of returns is not an equilibrium distribution. Some economists and finance theorists have mistakenly adopted and propagated the strange notion that random motion of returns defines "equilibrium." However, this disagrees with the requirement of time-translational invariance that in equilibrium no averages of any moment of the distribution can change with time. Random motion in the market is due to trading and the excess demand of unfilled limit orders prevents equilibrium at all or almost all times. Apparently, what many economists mean by "equilibrium" is more akin to assuming the EMH or the absence of arbitrage opportunities, which have nothing to do with vanishing excess demand in the market.

The only dynamically consistent definition of equilibrium is vanishing excess demand: if p denotes the price of an asset then excess demand $\varepsilon(p,t)$ is defined by $dp/dt = \varepsilon(p,t)$ including the case where the right-hand side is drift plus noise, as in stochastic dynamical models of the market. Bodie and Merton (1998) claim that vanishing excess demand is necessary for the CAPM, but we will see below that no such assumption comes into play during the derivation and would even cause *all* returns to vanish in the model.

The CAPM can be stated in the following way: Let R_0 denote the risk-free interest rate.

$$x_k = \ln(p_k(t + \Delta t)/p_k(t)) \tag{5.29}$$

is the fluctuating return on asset k where $p_k(t)$ is the price of the kth asset at time t. The total return x on the portfolio of n assets relative to the risk-free rate is given by

$$x - R_0 = \sum_{i=0}^{n} f_i(x_i - R_0) \tag{5.30}$$

where f_k is the fraction of the total budget that is bet on asset k. The CAPM minimizes the mean square fluctuation

$$\sigma^2 = \sum_{i,j} f_i f_j \langle (x_i - R_0)(x_j - R_0) \rangle = \sum_{i,j} f_i f_j \sigma_{ij} \tag{5.31}$$

subject to the constraints of fixed expected return R,

$$R - R_0 = \langle (x - R_0) \rangle = \sum_{i} f_i \langle (x_i - R_0) \rangle = \sum_{i} f_i (R_i - R_0) \tag{5.32}$$

and fixed normalization

5.4 The CAPM portfolio selection strategy

$$\sum_{i=0}^{n} f_i = 1 \tag{5.33}$$

where σ_{ij} is the correlation matrix

$$\sigma_{ij} = \langle (x_i - R_0)(x_j - R_0) \rangle \tag{5.34}$$

Following Varian (1992), we solve

$$\sigma_{ij} = \langle (x_i - R_0)(x_j - R_0) \rangle \tag{5.35}$$

for the fs where $\Delta R_e = R_e - R_0$ and R_e is the expected return of the "efficient portfolio," the portfolio constructed from fs that satisfy the condition (5.35). The expected return on asset k can be written as

$$\Delta R_k = \frac{\sigma_{ke}}{\sigma_{ee}} \Delta R_e = \beta_k \Delta R_e \tag{5.36}$$

where s^2 is the mean square fluctuation of the efficient portfolio, σ_{ke} is the correlation matrix element between the kth asset and the efficient portfolio, and ΔR_e is the "risk premium" for asset k.

The quantity β is interpreted as follows: $\beta = 1$ means the portfolio moves with the efficient portfolio, $\beta < 0$ indicates anticorrelation, and $\beta > 1$ means that the swings in the portfolio are greater than those of the efficient one. Small β indicates weak correlations but $\beta = 0$ doesn't signal statistical independence. Greater β also implies greater risk; to obtain a higher expected return you have to take on more risk. In the finance literature $\beta = 1$ is interpreted as reflecting moves with the market as a whole, but we will analyze and criticize this assumption below (in rating mutual funds, as on morningside.com, it is usually assumed that $\beta = 1$ corresponds to the market, or to a stock index). Contradicting the prediction of CAPM, studies show that portfolios with the highest βs usually yield lower returns historically than those with the lowest βs (Black *et al.*, 1972). This indicates that agents do not minimize risk as is assumed by the CAPM.

In formulating and deriving the CAPM above, nothing is assumed either about diversification or how to choose a winning portfolio. CAPM only advises us how to try to minimize the fluctuations in any arbitrarily chosen portfolio of n assets. The *a priori* chosen portfolio may or may not be well diversified relative to the market as a whole. It is allowed in the theory to consist entirely of a basket of losers. However, the qualitative conclusion that we can draw from the final result is that we should avoid a basket of losers by choosing assets that are anticorrelated with each other. In other words, although diversification is not necessarily or explicitly a *sine qua non*, we are advised by the outcome of the calculation to diversify in order to reduce risk. And on the other hand we are also taught that in order to expect large gains we should take on more risk. In other words, diversification is only one of two mutually exclusive messages gleaned from CAPM.

In the model, negative x represents a short position, positive x a long position. Large β implies both greater risk and larger expected return. Without larger expected return a trader will not likely place a bet to take on more risk. Negative returns R can and do occur systematically in market downturns, and in other bad bets.

In the finance literature the efficient portfolio is identified as the market as a whole. This is an untested assumption: without the required empirical analysis, there is no reason to believe that the entire Nasdaq or New York Exchange reflect the particular asset mix of an efficient portfolio, as if "the market" would behave as a CAPM risk-minimizing computer. Also, we will show below that Black–Scholes option pricing does not follow the CAPM strategy of risk minimization, but instead reflects a different strategy. In general, all that CAPM does is assume that n assets are chosen by any method or arbitrariness whatsoever. Given those n assets, CAPM shows how to minimize risk with return held fixed. The identification of the efficient portfolio as the market confuses together two separate definitions of efficiency: (1) the CAPM idea of an arbitrarily chosen portfolio with an asset mix that minimizes the risk, and (2) the EMH. The latter has nothing at all to do with portfolio selection.

Finance theorists distinguish systematic or market risk from diversifiable risk. The latter can be reduced, e.g. via CAPM, whereas we have no control over the former. The discussion that follows is an econophysics treatment of that subject.

Let's think of a vector f with entries $(f_1,\ldots,f_n)$ and a matrix Σ with elements σ_{kl}. The scalar product of f with σf is the mean square fluctuation

$$\sigma^2 = \tilde{f} \Sigma f \qquad (5.37)$$

If next we define a transformation U

$$w = Uf$$
$$\Lambda = U\Sigma\tilde{U} \qquad (5.38)$$

that diagonalizes Σ then we obtain

$$\sigma^2 = \sum_{k=1}^{n} w_k^2 \Lambda_k \qquad (5.39)$$

For many assets n in a well-diversified portfolio, studying the largest eigenvalue Λ_1 of the correlation matrix Σ has shown that that eigenvalue represents the market as a whole, and that clusters of eigenvalues represent sectors of the market like transportation, paper, autos, computers, etc. Here, we've ordered eigenvalues so that $\Lambda_1 \geq \Lambda_2 \geq \ldots \geq \Lambda_n$. In equation (5.39)

$$\sigma^2 = w_1^2 \Lambda_1 + \sum_{k=2}^{n} w_k^2 \Lambda_k \tag{5.40}$$

the first term represents so-called "nondiversifiable risk," risk due to the market as a whole, while the second term (the sum from 2 to n) represents risk that can be reduced by diversification. If we could assume that a vector component has order of magnitude $w_k = O(1/n)$ then we would arrive at the estimate

$$\sigma^2 \approx w_1^2 \Lambda_1^2 + \frac{\Lambda_k^2}{n} \tag{5.41}$$

which indicates that n must be very large in order to effectively get rid of diversifiable risk.

Let us consider a portfolio of two assets, e.g. a bond (asset #1) and the corresponding European call option (asset #2). For any two assets the solution for the CAPM portfolio can be written in the form

$$f_1/f_2 = (\sigma_{12}\Delta R_2 - \sigma_{22}\Delta R_1)/(\sigma_{12}\Delta R_1 - \sigma_{11}\Delta R_2) \tag{5.42}$$

Actually there are three assets in this model because a fraction f_0 can be invested in a risk-free asset, or may be borrowed in which case $f_0 < 0$. With only two assets, data analysis indicates that the largest eigenvalue of Λ apparently still represents the market as a whole, more or less (Laloux et al., 1999; Plerou et al., 1999). This means simply that the market systematically tends to drag the assets up or down with it.

5.5 Hedging with options

Fischer Black was motivated historically to try to use the CAPM for option pricing. His idea was simple: construct a two-asset portfolio consisting of a stock and the corresponding option. After introducing the idea of options and hedging we'll revisit Black's original idea. As is often the case, the original idea did not work as had been hoped, but a better idea for pricing options was found along the way.

Futures and options are examples of financial derivatives. A derivative is a bet contingent on the behavior of another financial asset, like a stock. An option is a contract that gives you the right but not the obligation to buy or sell an asset at a preselected price. The preselected price is called the strike price, K, and the deadline for exercising the option is called the expiration time T. An option to buy a financial asset is a call, an option to sell the asset is a put. A so-called "American option" can be exercised on or before its expiration time. A so-called "European option" can only be exercised at the strike time. These are only names having nothing to do with geography. An elementary description of options can be found in Bodie and Merton (1998), while Hull (1996) provides advanced details. Some familiarity with options is necessary in order to follow the text. For example, the reader should learn how to read and understand Figure 5.4.

We assume a so-called "frictionless" liquid market: we ignore all transaction fees, dividends, and taxes. We discuss only the so-called "European option" because it has mathematically the simplest forward-time initial condition, but has nothing geographically to do with Europe (in the context of the money supply, so-called "Eurodollars" also have nothing geographically to do with Europe).

Consider first a call. We want to know the value C of the call at a time $t < T$. C will depend on $(p(t), K, T-t)$ where $p(t)$ is the observed price at time t. In what follows $p(t)$ is assumed known. At $t = T$ we know that

$$C = \max[p(T) - K, 0] = (p(T) - K)\theta(p(T) - K) \tag{5.43}$$

where $p(T)$ is the price of the asset at expiration. Likewise, a put at exercise time T has the value

$$P = \max[K - p(T), 0] = (K - p(T))\theta(K - p(T)) \tag{5.44}$$

The main question is: what are the expected values of C and P at an earlier time $t < T$? The final price $p(T)$, unknown at time $t < T$, must be averaged and then discounted over the time interval $\Delta t = T - t$ at some rate r_d

$$C(p, K, T-t) = e^{-r_d(T-t)} \langle (p(T) - K)\vartheta(p(T) - K) \rangle \tag{5.45}$$

for the call. Clearly, this average is conditioned on observing p at time t so that the transition density for the price process is required. For the put,

$$P(p, K, T-t) = e^{-r_d(T-t)} \langle (K - p(T))\vartheta(K - p(T)) \rangle \tag{5.46}$$

Note that

$$C - P = e^{-r_d(T-t)}(\langle p(T) \rangle - K) = V - e^{-r_d(T-t)}K \tag{5.47}$$

where V is the expected asset price $p(T)$ at expiration, discounted back to time t at interest rate r_d where $r_0 \leq r_d$. The identity

$$C + e^{-r_d(T-t)}K = P + V \tag{5.48}$$

LIFFE EQUITY OPTIONS

Option		CALLS Apr	Jul	Oct	PUTS Apr	Jul	Oct
Alld Lyons	550	48	60	68	10	24	30
(*585)	600	22	34	44	34	50	55
ASDA	57	9	12½	14	4	7	9
(*61)	67	4	8	9	8½	11½	14
Brit Airways	280	27	33	39	9½	20	24
(*296)	300	17	24	29	18	30	34
SmKl Bchm A	460	30	47	53	17	26	52
(*470)	500	13	27	36	42	48	54
Boots	500	23	32	43	21	33	37
(*501)	550	6½	15	23	56	67	70
B P	260	16½	23	29	11	17	21
(*265)	280	8	15	19	22	28	34
British Steel	70	13	16½	19	3½	6½	8½
(*79)	80	8	12	14	7½	11½	13
Bass	600	31	48	61	24	38	45
(*604)	650	12	27	39	58	70	73
C & Wire	700	40	60	70	25	43	50
(*710)	750	18	35	46	56	73	80
Courtaulds	550	40	52	61	17	30	37
(*568)	600	16	29	38	45	60	64
Com Union	600	35	49	57	21	32	40
(*623)	650	11	26	34	53	61	70
Fisons	220	22	31	39	18	30	35
(*222)	240	14	22	31	31	43	48
GKN	460	27	36	45	21	28	34
(*472)	500	8½	20	28	48	53	57
Grand Met	420	38	53	59	11	21	27
(*439)	460	16	31	36	30	41	47
I.C.I	1100	53	82	92	50	68	88
(*1132)	1150	31	62	72	80	100	117
Kingfisher	550	34	48	53	21	38	45
(*559)	600	12	25	33	52	68	73
Ladbroke	200	16	25	29	18	26	32
(*202)	220	8	16	22	32	38	44
Land Secur	460	45	49	53	5	17	21
(*493)	500	17	25	31	20	38	41
M & S	330	18	25	34	12	20	24
(*333)	360	6	12	20	32	38	41
Sainsbury	550	43	54	63	12	24	29
(*577)	600	16	28	38	38	50	56
Shell Trans	550	32	44	50	12	19	26
(*576)	600	6	19	25	44	47	53
Storehouse	200	18	26	34	8	17	18
(*204)	220	11	17	22	21	25	28
Trafalgar	90	11	14	18	8	9	13
(*93)	100	6½	11	14	12	17	18
Utd Biscuits	360	18	25	33	19	25	29
(*366)	390	6	14	20	41	45	49
Unilever	1100	72	90	110	16	35	42
(*1149)	1150	39	60	82	42	56	63

Option		CALLS Feb	May	Aug	PUTS Feb	May	Aug
Brit Aero	280	23	40	53	16	34	46
(*287)	300	14	33	47	27	49	60

Option		CALLS Feb	May	Aug	PUTS Feb	May	Aug
BAA	750	41	61	71	6	16	31
(*786)	800	11	33	45	29	41	55
BAT Inds	950	45	58	74	8	37	46
(*982)	1000	16	33	50	30	64	73
BTR	550	22	30	38	5½	20	28
(*565)	600	3	10	19	38	53	55
Brit Telecom	420	10	23	29	7	15	24
(*420)	460	1	8	12½	40	41	50
Cadbury Sch	460	15	25	35	8	24	30
(*465)	500	3½	10	19	37	51	55
Eastern Elec	400	18	32	–	6	15	–
(*411)	430	5	–	–	24	–	–
Guinness	460	24	36	46	8	25	30
(*473)	500	6	19	28	33	48	56
GEC	300	9	19	23	7	13	21
(*301)	330	1	7	11	30	32	41
Hanson	260	7½	14½	18	5½	12	16½
(*262)	280	1	5½	9½	20	25	28
LASMO	160	8	18	22	9	18	23
(*161)	180	3	8½	15	24	33	36
Lucas Inds	140	15	23	26	4	11	16
(*151)	160	4	13	17	14	24	28
P. & O	550	24	42	52	15	40	52
(*564)	600	5½	21	33	45	72	82
Pilkington	100	7	15	18	6	12	16
(*102)	110	4	10	14	12	17	22
Prudential	300	24	30	34	2½	11	15
(*321)	330	6	13	19	14	26	31
R T Z	650	34	47	60	9	30	40
(*672)	700	10	25	39	36	59	69
Scot. & New	420	23	38	45	5	13	24
(*343)	460	4½	17	24	26	35	48
Tesco	240	22	27	30	2	8	12
(*257)	260	7	14	21	9	19	22
Thames Wtr	460	24	37	42	3½	12	22
(*479)	500	3½	16	20	25	32	45
Vodafone	390	18	34	43	8	21	28
(*398)	420	5½	19	29	27	36	45

Option		CALLS Mar	Jun	Sep	PUTS Mar	Jun	Sep
Abbey Nat	360	28	34	42	11	17	23
(*379)	390	11	19	27	27	33	38
Amstrad	20	5½	6½	7½	1½	1½	2½
(*24)	25	2½	3½	4½	3½	4	4½
Barclays	420	45	54	59	11	21	30
(*458)	460	20	33	38	33	41	51
Blue Circle	220	22	30	35	10	21	27
(*230)	240	12	19	27	22	33	39
British Gas	280	15½	19	23	6½	15	18
(*287)	300	6½	9½	14	18	27	30
Dixons	220	15	25	28	13	19	25
(*221)	240	8	17	21	26	31	38
Eurotunnel	420	38	55	70	20	35	45
(*435)	460	18	37	50	45	57	67

Option		CALLS Mar	Jun	Sep	PUTS Mar	Jun	Sep
Glaxo	650	38	60	82	29	42	52
(*664)	700	17	38	58	60	72	80
Hillsdown	140	17	22	24	6	15	20
(*147)	160	6	12	16	18	27	31
Lonrho	70	9	12	15	5	8½	11
(*75)	80	4½	8	11½	11	14	17
HSBC 75p shs	550	45	57	70	17	34	46
(*576)	600	18	32	48	44	60	70
Natl Power	280	24	32	36	6	11	18
(*294)	300	11	21	25	15	20	27
Reuters	1400	69	108	135	47	78	100
(*1426)	1450	40	82	113	78	103	122
R. Royce	130	11	15	19	9	14	17
(*134)	140	6	11	16	16	20	23
Scot Power	200	20	25	28	2	5	10½
(*216)	220	6	13	17	9	13½	22
Sears	100	9	11	15	5	8½	11
(*104)	110	4	7	9½	10	14	16
Forte	180	15	20	25	10	19	24
(*186)	200	7	13	17	21	32	36
Thorn EMI	800	61	81	90	10	22	39
(*843)	850	28	50	60	29	43	66
TSB	160	19	23	27	5	8½	12
(*176)	180	6½	13	17	15	18	22
Vaal Reefs	30	6	6	6½	3	4	4½
(*$34)	35	2	4	5½	5½	6	8
Wellcome	850	50	75	100	28	48	63
(*866)	900	26	52	75	57	77	92

EURO FT-SE INDEX (*2872)

	2675	2725	2775	2825	2875	2925	2975	3025
CALLS								
Feb	205	155	108	68	36	16	5	2
Mar	215	170	129	92	61	38	22	11
Apr	–	188	–	115	–	61	–	28
Jun	–	217	–	152	–	100	–	60
Sep	–	252	–	188	–	129	–	–
PUTS								
Feb	1½	3½	7	15	34	65	104	151
Mar	9	12	21	34	53	81	114	153
Apr	–	26	–	51	–	97	–	163
Jun	–	50	–	80	–	125	–	185
Sep	–	–	120					

FT-SE INDEX (*2872)

	2650	2700	2750	2800	2850	2900	2950	3000
CALLS								
Feb	233	183	135	89	50	24	9	3
Mar	243	197	153	113	79	51	31	17
Apr	254	210	170	131	99	73	51	36
May	266	226	186	151	121	93	72	52
Jun	–	235	–	164	–	107	–	67
Dec †	–	–	–	235	–	187	–	130
PUTS								
Feb	1½	2½	4½	10	23	47	85	131
Mar	8	12	17	29	46	69	100	138
Apr	16	22	31	44	63	86	116	151
May	26	35	45	61	80	105	133	166
Jun	–	40	–	70	–	113	–	175
Dec †	–	90	–	125	–	185	–	–

February 3 Total Contracts 31,257
Calls 21,861 Puts 9,396
FT-SE Index Calls 7,946 Puts 4,410
Euro FT-SE Calls 816 Puts 278
*Underlying security price.   † Long dated expiry mths
Premiums shown are based on closing offer prices.

Figure 5.4 Table of option prices from the February 4, 1993 *Financial Times*. From Wilmott, Howison, and DeWynne (1995), Figure 1.1.

is called put-call parity, and provides a starting point for discussing so-called synthetic options. That is, we can simulate puts and calls by holding some combination of an asset and money market.

Suppose first that we finance the trading by holding an amount of money $M_0 = -e^{-r_d(T-t)}K$ in a risk-free fund like a money market, so that $r_d = r_0$ where r_0 is the risk-free interest rate, and also invest in one call. The value of the portfolio is

$$\Pi = C + e^{-r_0(T-t)}K \qquad (5.49)$$

This result synthesizes a portfolio of exactly the same value made up of one put and one share of stock (or one bond)

$$\Pi = V + P \qquad (5.50)$$

and vice versa. Furthermore, a call can be synthesized by buying a share of stock (taking on risk) plus a put (buying risky insurance[2]),

$$C = P + V - e^{-r_0(T-t)}K \qquad (5.51)$$

while borrowing an amount M_0 (so-called risk-free leverage).

In all of the above discussion we are assuming that fluctuations in asset and option prices are small, otherwise we cannot expect mean values to be applicable. In other words, we must expect the predictions above to fail in a market crash when liquidity dries up. Option pricing via calculation of expectation values can only work during normal trading when there is adequate liquidity. LTCM failed because they continued to place "normal" bets against the market while the market was going against them massively (Dunbar, 2000).

5.6 Stock shares as options on a firm's assets

We reproduce in part here an argument from the original paper by Black and Scholes (1973) that starts with the same formula as the M & M argument, $p = B + S$ where p is the current market estimate of the value of a firm, B is debt owed to bondholders, and S is the current net value of all shares of stock outstanding. Black and Scholes noticed that their option-pricing formula can be applied to this valuation $p = B + S$ of a firm. This may sound far-fetched at first sight, but the main point to keep in mind in what follows is that

---

[2] This form of insurance is risky because it's not guaranteed to pay off, in comparison with the usual case of life, medical, or car insurance.

5.6 Stock shares as options on a firm's assets

bondholders have first call on the firm's assets. Unless the bondholders can be paid in full the shareholders get nothing.

The net shareholder value at time t is given by $S = N_s p_s$ where N_s is the number of shares of stock outstanding at price p_s. To keep the math simple we assume in what follows that no new shares are issued and that all bonds were issued at a single time t_0 and are scheduled to be repaid with all dividends owed at a single time T (this is a mathematical simplification akin to the assumption of a European option). Assume also that the stock pays no dividend. With N_s constant the dynamics of equity S are the same as the dynamics of stock price p_s. Effectively, the bondholders have first call on the firm's assets. At time T the amount owed by the firm to the bondholders is $B'(T) = B(T) + D$, where $B(T)$ is the amount borrowed at time t_0 and D is the total interest owed on the bonds. Note that the quantity $B'(T)$ is mathematically analogous to the strike price K in the last section on options: the stock share is worth something if $p(T) > B'(T)$, but is otherwise worthless. At expiration of the bonds, the shareholders' equity, the value of all shares, is then

$$S(T) = \max(p(T) - B'(T), 0) \quad (5.52)$$

Therefore, at time $t < T$ we can identify the expected value of the equity as

$$S(p, B'(T), T - t) = e^{-r_d(T-t)} \langle \max(p(T) - B'(T), 0) \rangle \quad (5.53)$$

showing that the net value of the stock shares S can be viewed formally for $t < T$ as an option on the firm's assets. Black and Scholes first pointed this out. This is a very beautiful argument that shows, in contrast with the famous 1970s-style brokerage house advertisement "Own a Piece of America," a stock shareholder owns nothing but an option on future equity, so long as there is corporate debt outstanding. An option risks loss of capital via the market turning against the bet; a money market account risks loss of capital via inflation. But as at least one famous trader has stated, in times of uncertainty liquidity is king.

Of course, we have formally treated the bondholder debt as if it would be paid at a definite time T, which is not realistic, but this is only an unimportant detail that can be corrected by a much more complicated mathematical formulation. That is, we have treated shareholder equity as a European option, mathematically the simplest kind of option. The idea here is to illustrate an idea, not to provide a calculational recipe.

The idea of a stock as an option on a company's assets is theoretically appealing: a stockholder owns no physical asset, no buildings, no equipment, etc., at $t < T$ (all debt is paid hypothetically at time T), and will own real assets like plant, machinery, etc. at $t > T$ if and only if there is anything left over after the bondholders have been paid in full. The Black–Scholes explanation of shareholder value reminds us superficially of the idea of book or replacement value mentioned in Section 4.2, which is based on the idea that the value of a stock share is determined by the value of a firm's net real and financial assets after all debt obligations have been subtracted. However, in a bubble the equity S can be inflated, and S is anyway generally much larger than book or replacement value in a typical market. That S can be inflated is in qualitative agreement with M & M, that shares are bought based on future expectations of equity growth ΔS. In this formal picture we only know the dynamics of $p(t)$ through the dynamics of B and S. The valuation of a firm on the basis of $p = B + S$ is not supported by trading the firm itself, because even in a liquid equity market Exxon, Intel, and other companies do not change hands very often. Thinking of $p = B + S$, we see that if the firm's bonds and shares are liquid in daily trading, then that is as close to the notion of liquidity of the firm as one can get.

The "air" never came out of the market after the dot.com bubble popped in 2001. Far too much money was/is in circulation for a deflation to occur (the money supply is discussed in Chapter 9). See Baruch's autobiography (1957) for his account of how he bought railroad stocks at greatly lowered costs after the onset of the liquidity crunch (the Great Depression) following the 1929 market crash.

5.7 The Black–Scholes model

The Black–Scholes model can be derived in all detail from a special portfolio called the delta hedge (Black and Scholes, 1973). Let $w(p,t)$ denote the option price. Consider a portfolio short one call option and long Δ shares of stock. "Long" means that the asset is purchased, "short" means that it is sold. If we choose $\Delta = w'$ then the portfolio is instantaneously risk-free. To see this, we calculate the portfolio's value at time t

$$\Pi = -w + \Delta p \qquad (5.54)$$

Using the Gaussian returns model we obtain the portfolio's rate of return (after using $dB^2 = dt$)

$$\begin{aligned}\frac{d\Pi}{\Pi dt} &= (-dw + \Delta dp)/\Pi dt \\ &= (-\dot{w}\Delta t - w' dp - w'' \sigma_1^2 p^2/2 + \Delta dp)/\Pi dt\end{aligned} \qquad (5.55)$$

Here, we have held the fraction Δ of shares constant during dt because this is what the hypothetical trader must do. If we choose $\Delta = w'$ then the portfolio

5.7 The Black–Scholes model

has a deterministic rate of return $d\Pi/\Pi dt = r$. In this special case, called the delta hedge portfolio, we obtain

$$\frac{d\Pi}{\Pi dt} = (-\dot{w}dt - w''\sigma_1^2 p^2/2)/(-w + w'p)dt = r \qquad (5.56)$$

where the portfolio return r does not fluctuate randomly to $O(dt)$ and must be determined or chosen. In principle r may depend on (p,t). The cancellation of the random term $w'dp$ in the numerator of (5.56) means that the portfolio is *instantaneously* risk-free: the mean square fluctuation of the rate of return $d\Pi/\Pi dt$ vanishes to $O(dt)$,

$$\left\langle \left(\frac{d\Pi}{\Pi dt} - r\right)^2 \right\rangle = 0 \qquad (5.57)$$

but not to higher order. This is easy to see. With $w(p,t)$ deterministic the finite change $\Delta\Pi = -\Delta w + w' \bullet \Delta p$ fluctuates over a finite time interval due to Δp. This makes the real portfolio risky because continuous time portfolio rebalancing over infinitesimal time intervals dt is impossible in reality.

The delta hedge portfolio is therefore not globally risk-free like a CD where the mean square fluctuation vanishes for all finite times Δt. To maintain the portfolio balance as the observed asset price p changes while t increases toward expiration, the instantaneously risk-free portfolio must continually be updated. This is because p changes and both w and w' change with t and p. Updating the portfolio frequently is called "dynamic rebalancing." Therefore the portfolio is risky over *finite* time intervals Δt, which makes sense: trading stocks and options, in any combination, is a very risky business, as any trader can tell you.

The standard assumption among finance theorists is that $r = r_0$ is the risk-free rate of interest. Setting $r = r_0$ means that one assumes that the hedge portfolio is perfectly equivalent to a money market deposit, which is wrong. Note, however, that (5.56) holds for any value of r. The theory does not pick out a special value for the interest rate r of the hedge portfolio.

Finally, with $r = d\Pi/\Pi dt$ in (5.56) we obtain the famous Black–Scholes pde

$$rw = r\dot{w} + rpw' + \frac{1}{2}\sigma_1^2 p^2 w'' \qquad (5.58)$$

a backward-in-time diffusion equation that revolutionized finance. The initial condition is specified at a forward time, the strike time T, and the equation diffuses backward in time from the initial condition to predict the option price $w(p,t)$ corresponding to the observed asset price p at time t. For a call, for example, the initial condition at expiration is given by (5.43).

Black, Scholes, and Merton were not the first to derive option-pricing equations; Samuelson and others had made option-pricing models. Black, Scholes, and Merton were the first to derive a falsifiable option-pricing pde by using only observable quantities. Long before their famous discovery, Black was an undergraduate physics student, Scholes was an economist with a lifelong interest in the stock market, and Merton was a racing car enthusiast/mechanic who played the stock market as a student. Samuelson revived Ito calculus and Merton, a strong mathematician, put it to work in finance theory.

In their very beautifully written original 1973 paper Black and Scholes produced two separate proofs of the pde (5.58), one from the delta hedge and the other via CAPM. Black (1989) has explained that the CAPM provided his original motivation to derive an option-pricing theory. We will show next that CAPM does not lead to (5.58) but instead assumes a different risk-reduction strategy, so that the original Black–Scholes paper contains an error.

Strangely enough, Steele (2000) apparently was aware of that mistake but wrote a weak excuse for Black and Scholes instead of presenting a clarification. Worse, he wrote as if the formula justified the wrong derivation, which is bad advice. Science is not like religion where humans are canonized as perfection with human errors erased from history. We should learn from mistakes instead of sweeping them under the rug.

5.8 The CAPM option pricing strategy

In what follows we consider the CAPM for two assets, a stock or bond with rate of return R_1, and a corresponding option with rate of return R_2. Assuming lognormal asset pricing (5.54) the average return on the option is given by the sde for w as

$$\mathrm{d}w = (\dot{w} + R_1 p w' + \sigma_1^2 p^2 w''/2)\mathrm{d}t + pw'\sigma_1 \mathrm{d}B \qquad (5.59)$$

where we've used $\mathrm{d}B^2 = \mathrm{d}t$. This yields an instantaneous rate of return on the option

$$x_2 = \frac{\mathrm{d}w}{w\mathrm{d}t} = \frac{\dot{w}}{w} + \frac{pw'}{w}R_1 + \frac{1}{2}\sigma_1^2 p^2 \frac{w''}{w} + \frac{pw'}{w}\sigma_1^2 \frac{\mathrm{d}B}{\mathrm{d}t} \qquad (5.60)$$

where $\mathrm{d}B/\mathrm{d}t$ is white noise. From CAPM we have

$$R_2 = R_0 + \beta_2 \Delta R_e \qquad (5.61)$$

for the average return. The average return on the stock is given from CAPM by

5.8 The CAPM option pricing strategy

$$R_1 = R_0 + \beta_1 \Delta R_e \tag{5.62}$$

and the instantaneous return rate is $x_2 = dp/pdt = R_1 \sigma_1 dB/dt$. According to the original Nobel Prize-winning 1973 Black–Scholes paper we should be able to prove that

$$\beta_2 = \frac{pw'}{w}\beta_1 \tag{5.63}$$

Were this the case then we would get a cancellation of the two β terms in (5.64) below:

$$\begin{aligned}R_2 = R_0 + \beta_2 \Delta R_e &= \frac{\dot{w}}{w} + \frac{pw'}{w}R_1 + \frac{1}{2}\sigma_1^2 p^2 \frac{w''}{w} \\ &= \frac{\dot{w}}{w} + \frac{pw'}{w}R_0 + \frac{pw'}{w}\beta_1 \Delta R_e + \frac{1}{2}\sigma_1^2 p^2 \frac{w''}{w}\end{aligned} \tag{5.64}$$

leaving us with risk-free rate of return R_0 and the Black–Scholes option-pricing pde (5.63). We show next that this result would only follow from a circular argument and is wrong: the two β terms do not cancel each other.

From the sde (5.59) for w the fluctuating option price change over a finite time interval Δt is given by the stochastic integral equation

$$\Delta w = \int_t^{t+\Delta t} (\dot{w} + w'R_1 p + \frac{1}{2}w''\sigma_1^2 p^2)dt + \sigma_1(w'p) \bullet \Delta B \tag{5.65}$$

where the dot in the last term denotes the Ito product. In what follows we assume sufficiently small time intervals Δt to make the small returns approximation whereby $\ln(w(t+\Delta t)/w(t)) \approx \Delta w/w$ and $\ln(p(t+\Delta t)/p(t)) \approx \Delta p/p$. In the small returns approximation (local solution of (5.65))

$$\Delta w \approx (\dot{w} + w'R_1 p + \frac{1}{2}w''\sigma_1^2 p^2)\Delta t + \sigma_1 w'p\Delta B \tag{5.66}$$

We can use this to calculate the fluctuating option return $x_2 \approx \Delta w/w\Delta t$ at short times. With $x_1 \approx \Delta p/p\Delta t$ denoting the short time approximation to the asset return, we obtain

$$x_2 - R_0 \approx \frac{1}{w}(\dot{w} + \frac{\sigma_1^2 p^2 w''}{2} + R_0 pw' - R_0 w) + \frac{pw'}{w}(x_1 - R_1) \tag{5.67}$$

Taking the average would yield (5.63) *if we were to assume* that the Black–Scholes pde (5.58) holds, but we are trying to *derive* (5.58), not assume it. Therefore, taking the average yields

$$\beta_2 \gg \frac{1}{w\Delta R_e}\left(\frac{\partial^2 w}{\partial t^2} + \frac{\sigma_1^2 p^2 w''}{2} + R_0 pw' - R_0 w\right) + \frac{pw'}{w}\beta_1 \qquad (5.68)$$

which is true but does not reduce to (5.58), in contrast with the claim made by Black and Scholes. Equation (5.58) is in fact impossible to derive without making a circular argument. Within the context of CAPM one certainly cannot use (5.63) in the CAPM model.

To see that we cannot assume (5.58) just calculate the ratio f_2/f_1 invested by our hypothetical CAPM risk-minimizing agent. Here, we need the correlation matrix for Gaussian returns only to leading order in Δt:

$$\sigma_{11} \approx \sigma_1^2/\Delta t \qquad (5.69)$$

$$\sigma_{12} \approx \frac{pw'}{w}\sigma_{11} \qquad (5.70)$$

and

$$\sigma_{22} \approx \left(\frac{pw'}{w}\right)^2 \sigma_{11} \qquad (5.71)$$

The variance of the portfolio vanishes to lowest order as with the delta hedge, but it is also easy to show that to leading order in Δt

$$f_1 \propto (\beta_1 pw'/w - \beta_2)pw'/w \qquad (5.72)$$

and

$$f_2 \propto (\beta_2 - \beta_1 pw'/w) \qquad (5.73)$$

so that it is impossible that the Black–Scholes assumption (5.68) could be satisfied. Note that the ratio f_1/f_2 is exactly the same as for the delta hedge.

That CAPM is not an equilibrium model is exhibited explicitly by the time dependence of the terms in the averages used.

The CAPM does not predict either the same option-pricing equation as does the delta hedge. Furthermore, if traders actually use the delta hedge in option pricing then this means that agents do not trade in a way that minimizes the variance via CAPM. The CAPM and the delta hedge do not try to reduce risk in exactly the same way. In the delta hedge the main fluctuating terms are removed directly from the portfolio return, thereby lowering the expected return. In CAPM, nothing is subtracted from the return in forming the portfolio and the idea there is not only diversification but also increased expected return through increased risk. In other words, the delta hedge and CAPM attempt to minimize risk in two entirely different ways: the delta hedge attempts to eliminate risk altogether whereas in CAPM one acknowledges that higher risk is required for higher expected return. We see now that the way that options are priced is strategy dependent, which is closer to the idea that psychology plays a role in trading.

The CAPM option-pricing equation depends on the expected returns for both stock and option,

$$R_2 w = \dot{w} + pw'R_1 + \frac{1}{2}\sigma_1^2 p^2 w'' \tag{5.74}$$

and so differs from the original Black–Scholes equation (5.63) of the delta hedge strategy. There is no such thing as a universal option-pricing equation independent of the chosen strategy, even if that strategy is reflected in this era by the market. Economics is not like physics (non-thinking nature), but depends on human choices and expectations. This can easily be forgotten by financial engineers under pressure to invent ever-new derivatives in order to circumvent regulations and make new sales.

5.9 Backward-time diffusion: solving the Black–Scholes pde

Next, we show that it is very simple to use the Green function method from physics to solve the Black–Scholes pde, which is a simple, linear, backward-in-time diffusion equation. This approach is much more transparent than the standard one found in finance texts.

Consider the simplest diffusion equation

$$\frac{\partial f}{\partial t} = D \frac{\partial^2 f}{\partial x^2} \tag{5.75}$$

with $D > 0$ a constant. Solutions exist only forward in time; the time evolution operator

$$U(t) = e^{tD\frac{\partial^2}{\partial x^2}} \tag{5.76}$$

has no inverse. The solutions

$$f(x,t) = U(t)f(x,0) = f(x,0) + tD\frac{\partial f(x,0)}{\partial x} + \cdots + \frac{(tD)^n}{n!}\frac{\partial^n f(x,0)}{\partial x^n} + \cdots \tag{5.77}$$

form a semi-group. The infinite series (5.77) is equivalent to the integral operator

$$f(x,t) = \int_{-\infty}^{\infty} g(x,t|z,0)f(z,0)\,dz \tag{5.78}$$

where g is the Green function of (5.78). That there is no inverse of (5.76) corresponds to the nonexistence of the integral (5.78) if t is negative.

Consider next the diffusion equation (Sneddon, 1957)

$$\frac{\partial f}{\partial t} = -D\frac{\partial^2 f}{\partial x^2} \tag{5.79}$$

It follows that solutions exist only backward in time, with t starting at t_0 and decreasing. The Green function for (5.79) is given by

$$g(x,t|x_0,t_0) = \frac{1}{\sqrt{4\pi D(t_0-t)}} e^{-\frac{(x-x_0)^2}{4D(t_0-t)}} \tag{5.80}$$

With arbitrary initial data $f(x,t_0)$ specified forward in time, the solution of (5.79) is for $t \leq t_0$ given by

$$f(x,t) = \int_{-\infty}^{\infty} g(x,t|z,t_0)f(z,t_0)dz \tag{5.81}$$

where the Green function is the transition density for the Markov process. We can rewrite the equations as forward in time by making the transformation $\Delta t = t_0 - t$ so that (5.79) and (5.80) become

$$\frac{\partial f}{\partial \Delta t} = D\frac{\partial^2 f}{\partial x^2} \tag{5.82}$$

and

$$g(x,t|x_0,t_0) = \frac{1}{\sqrt{4\pi D \Delta t}} e^{-\frac{(x-x_0)^2}{4D\Delta t}} \tag{5.83}$$

with Δt increasing as t decreases.

We can solve the Black–Scholes pde as follows. Starting with the Black–Scholes pde (5.58) and transforming to returns x

$$ru = \dot{u} + r'u' + \frac{1}{2}\sigma_1^2 u'' \tag{5.84}$$

where $u(x,t) = w(p,t)$ transforms like price, not density. We next make the time-transformation $w = ve^{rt}$ so that

$$0 = \dot{v} + r'v + \frac{1}{2}\sigma^2 v'' \tag{5.85}$$

The Green function for this equation is the Gaussian

5.9 Backward-time diffusion

$$g(x - r'(T-t), T-t) = \frac{1}{\sigma_1\sqrt{4\pi^2(T-t)}} e^{-\frac{(x-r'(T-t))^2}{2\sigma_1^2(T-t)}} \tag{5.86}$$

and the forward-time initial condition for a call at time T is

$$\begin{aligned} v(x,T) &= e^{-rT}(pe^x - K), \quad x < 0 \\ v(x,T) &= 0, x > 0 \end{aligned} \tag{5.87}$$

so that the call has the value

$$C(K,p,T-t) = e^{-r(T-t)} p \int_{\ln K/p}^{\infty} g(x - r'(T-t), T-t) e^x dx$$

$$- e^{-r(T-t)} K \tag{5.88}$$

$$\int_{\ln K/p}^{\infty} g(x - r'(T-t), T-t) dx$$

The reader can write down the corresponding formula for a put. However, in the transformation back to price in (5.88), the Green function, a transition density, transforms like a density. That is, the option price C and the transition density in (5.88) obey completely different coordinate transformation rules.

By completing the square in the exponent of the first integral in (5.88) and then transforming variables in both integrals, we can transform equation (5.88) into the standard textbook form (Hull, 1997), convenient for numerical calculation:

$$C(K,p,T-t) = pN(d_1) - Ke^{-r\Delta t}N(d_2) \tag{5.89}$$

where

$$N(d) = \frac{1}{\sqrt{2\pi}} \int_{-\infty}^{d} e^{-y^2/2} dy \tag{5.90}$$

with

$$d_1 = \frac{\ln p/K + (r + \sigma_1^2/2)\Delta t}{\sigma_1\sqrt{\Delta t}} \tag{5.91}$$

and

$$d_2 = \frac{\ln p/K + (r - \sigma_1^2/2)\Delta t}{\sigma_1\sqrt{\Delta t}} \tag{5.92}$$

Finally, to complete the picture, Black and Scholes, following the theorists Modigliani and Miller, assumed the no-arbitrage condition. Because the portfolio is instantaneously risk-free they chose $r = r_0$, the bank interest rate. Again, the idea is to predict a hypothetically fair option price, then use it as benchmark to look for "mispricings" to trade on. See Wilmott (1995) for solutions of the Black–Scholes pde for the different boundary conditions of interest in financial engineering.

5.10 Enron 2002

The collapse of Enron can be discussed in the context of the Gambler's Ruin and the M & M theorem.

Enron (Bryce and Ivins, 2002) started by owning real assets in the form of gas pipelines, but became a so-called New Economy company during the 1990s based on the belief that derivatives trading, not assets, paves the way to great wealth acquired fast. This was during the era of widespread belief in reliable applicability of mathematical modeling of derivatives, and "equilibrium" markets, before the collapse of LTCM. At the time of its collapse, Enron was building the largest derivatives trading floor in the world.

Compared with other market players, Enron's Value at Risk (VaR, Jorion, 1997) and trading-risk analytics were "advanced," but were certainly not "fool-proof." Enron's VaR model was a modified Heath–Jarrow–Morton model utilizing numerous inputs (other than the standard price/volatility/position) including correlations between individual "curves" as well as clustered regional correlations, factor loadings (statistically calculated potential stress scenarios for the forward price curves), "jump factors" for power price spikes, etc. Component VaR was employed to identify VaR contributors and mitigators, and Extreme Value Theory[3] was used to measure potential fat-tail events. However, about 90% of the employees in "Risk Management" and virtually all of the traders could not list, let alone explain, the inputs into Enron's VaR model.[4]

A severe weakness is that Enron tried to price derivatives in nonliquid markets. This means that inadequate market returns or price histograms were used to try to price derivatives and assess risk. VaR requires good statistics for the estimation of the likelihood of extreme events, and so with an inadequate histogram the probability of an extreme event cannot be meaningfully estimated. Enron even wanted to price options for gas stored in the ground,

---

[3] See Sornette (1998) and Dacorogna *et al.* (2001) for definitions of Extreme Value Theory.
[4] The information in this paragraph was provided by a former Enron risk management researcher who prefers to remain anonymous.

an illiquid market for which price statistics could only be invented.[5] Some information about Enron's derivatives trading was reported in the article www.nytimes.com/2002/12/12/business/12ENER.html?pagewanted=1.

But how could Enron "manufacture" paper profits, without corresponding cash flow, for so long and remain undetected? The main accounting trick that allowed Enron to report false profits, driving up the price of its stock and providing enormous rewards to its deal-makers, was "mark to market" accounting. Under that method, future projected profits over a long time interval are allowed to be declared as current profit even though no real profit has been made, even though there is no positive cash flow. In other words, firms are allowed to announce to shareholders that profits have been made when no profit exists. Enron's globally respected accounting firm helped by signing off on the auditing reports, in spite of the fact that the auditing provided so little real information about Enron's financial status. At the same time, major investment houses that also profited from investment banking deals with Enron touted the stock.

Another misleading use of mark to market accounting is as follows: like many big businesses (Intel, GE, etc.), Enron owned stock in dot.com outfits that later collapsed in and after winter 2000, after never having shown a profit. When the stock of one such company, Rhythms NetConnections, went up significantly, Enron declared a corresponding profit on its books without having sold the stock. When the stock price later plummeted Enron simply hid the loss by transferring the holding into one of its spinoff companies. Within that spinoff, Enrons' supposed "hedge" against the risk was its own stock.

The use of mark to market accounting as a way of inflating profit sheets surely should be outlawed,[6] but such regulations fly in the face of the widespread belief in the infallibility of "the market mechanism." Shareholders should at the very least be made fully aware in quarterly reports of all derivatives positions held by a firm, and how big a fraction of a market those derivatives represent (this would help to expose potential liquidity problems under selling pressure). Ordinary taxpayers in the USA are not permitted to declare as profits or losses unrealized stock price changes. As Black and Scholes made clear, a stock is not an asset, it is merely an option on an asset. Real assets (money in the bank, plant and equipment, etc.), not unexercised options, should be the basis for deciding profits/losses and taxation. In addition, accounting rules should be changed to make it extremely difficult

---

[5] Private conversation with Enron modelers in 2000.
[6] It would be a good idea to mark liquid derivatives positions to market to show investors the level of risk. Illiquid derivatives positions cannot be marked to market in any empirically meaningful way, however.

for a firm to hide its potential losses on bets placed on other firms: all holdings should be declared in quarterly reports in a way that makes clear what are real assets and what are risky bets.

Let us now revisit the M & M theorem. Recall that it teaches that to a first approximation in the valuation of a business $p = B + S$ the ratio B/S of debt to equity doesn't matter. However, Enron provides us with examples where the amount of debt does matter. If a company books profits through buying another company, but those earnings gains are not enough to pay off the loan, then debt certainly matters. With personal debt, debt to equity matters since one can go bankrupt by taking on too much debt. The entire M & M discussion is based on the small returns approximation $E = <\Delta p> \approx p\Delta t$, but this fails for big changes in p. The discussion is therefore incomplete and can't be extrapolated to extreme cases where bankruptcy is possible. So the ratio B/S in $p = B + S$ does matter in reality, meaning that something important is hidden in the future expectations E and ignored within the M & M theorem.

Enron made a name for itself in electricity derivatives after successfully lobbying for the deregulation of the California market. The manipulations that were successfully made by options traders in those markets are now well documented. Of course, one can ask: why should consumers want deregulated electricity or water markets anyway? Deregulation lowered telephone costs, both in the USA and western Europe, but electricity and water are very different. Far from being an information technology, both require the expensive transport of energy over long distances, where dissipation during transport plays a big role in the cost. So far, in deregulated electricity and water markets, there is no evidence that the lowering of consumer costs outweighs the risk of having firms play games trying to make big wins by trading options on those services. The negative effects on consumers in California and Buenos Aires do not argue in favor of deregulation of electricity and water.

It's too easy to conclude that Adam Smith qualitatively extrapolated simple ideas of mechanical friction to conclude that supply and demand is self-stabilizing, analogous to the way that the laws of mechanics lead to a terminal speed for a ball falling through the air. But Smith, a Calvinist with strong moral principles, was not so naive. He asserted that moral restraint would be necessary for free markets to function correctly (in stable fashion). Smith was perceptive, but his hopes have not been realized. That moral restraint alone is inadequate to stabilize free markets is illustrated in Chapter 9, in the context of the Dollar under the gold standard.

6
Scaling, pair correlations, and conditional densities

We've covered the basic required math in Chapter 3, and have introduced the reader to the most basic ideas of financial markets in Chapter 4. Scaling is widely assumed in econophysics; the questions for us are simple: (i) what does scaling imply, and (ii) does it really occur (Chapter 7)? In this chapter we explicitly construct scaling models where one class violates the EMH and the other class satisfies it. We also determine whether scaling, when it occurs, is reflected in transition densities and pair correlations.

6.1 Hurst exponent scaling

We now begin to discuss two completely unrelated topics that are often confused together in the literature: scaling and long time correlations. Scaling with a Hurst exponent $H \neq 1/2$ is often misinterpreted as implying the long time autocorrelations of fractional Brownian motion (fBm). We'll show that scaling has nothing to do with long time correlations: when scaling occurs, then it's restricted to one-point densities and one-point densities tell us nothing about correlations. We'll show in the end that transition densities and pair correlations generally cannot scale even if certain random trajectories do scale. In other words, and in contrast with the statistical physics of order–disorder transitions, scaling does not reflect dynamics at all!

A stochastic process $x(t)$ is said to scale with Hurst exponent H if

$$x(t) = t^H x(1) \tag{6.1}$$

where by equality we mean equality "in distribution" (Embrechts and Maejima, 2002). We next define what that means for one class of processes. Clearly, $x(0) = 0$ *is necessary for random trajectories that scale*.

Consider next simple averages of a dynamical variable $A(x,t)$. Simple averages require only the one-point density,

$$\langle A(t)\rangle = \int_{-\infty}^{\infty} A(x,t)f_1(x,t)dx \tag{6.2}$$

From (6.1), the moments of x must obey

$$\langle x^n(t)\rangle = t^{nH}\langle x^n(1)\rangle = c_n t^{nH} \tag{6.3}$$

Combining this with

$$\langle x^n(t)\rangle = \int x^n f(x,t)dx \tag{6.4}$$

we obtain

$$f(x,t) = t^{-H} F(u) \tag{6.5}$$

where the scaling variable is $u = x/t^H$. This predicts a data collapse $F(u) = t^H f(x,t)$ that could be tested empirically, *if densities can be extracted reliably from empirical data* (see Chapter 7).

We'll see below that scaling generally requires a drift coefficient $R = 0$, so that with $x(0) = 0$ the unconditioned average of x vanishes. The variance is then simply

$$\sigma^2 = \langle x^2(t)\rangle = \langle x^2(1)\rangle t^{2H} \tag{6.6}$$

This explains what is meant by Hurst exponent scaling, and also specifies what's meant by that (6.1) holds "in distribution," namely, that the one-point density scales.

The Wiener process is an example of a Markov process that scales. One sees this from the one-point Gaussian density $f_1(B,t) = t^{-1/2} F(B/t^{1/2})$ where $F(u) = exp(-u^2/2)/(2\pi)^{1/2}$. We conclude that $B(t) = t^{1/2} B(1)$ with $B(0) = 0$. The two-point density obeys $p_2(B,t|B',t') = p_2(B-B', t-t'|0,0) = f_1(\Delta B, \Delta t)$ due to stationarity of the increments. Note that the transition density scales with the time lag $T = t - t'$ only because of time-translational invariance. The pair correlations are given by $\langle B(t+T)B(t)\rangle = \langle B^2(t)\rangle = t$ and do not scale with exponent $H = 1/2$ in time scales t and $t + T$.

Even if scaling occurs, it's broken by arbitrary choices of initial conditions $x(t_0) \neq 0$. A more precise way to say this is that the only density that may scale is $f_1(x,t) = p_2(x,t|0,0)$. We make this claim rigorous below, where we show that $p_2(y, t+T|x,t)$ generally does not scale.

Our scaling discussion above assumes finite moments of all orders. Both Levy and Markov processes can generate "fat tails," $f_1(x,t) \approx |x|^{-\mu}$ when $|x| \gg 1$, where moments $\langle x^n(t)\rangle$ blow up after some finite value of n. In that case one would have to discuss scaling using the one-point density directly (Scalas et al., 2000) without any appeal to the moments.

Nothing is implied about correlations and/or dynamics by Hurst exponent scaling. We'll exhibit this below by presenting both Markov and strongly correlated processes that scale with the same Hurst exponent H, and generate exactly the same one-point density. *This tells us that neither one-point densities, nor diffusion equations for one-point densities, imply a specific underlying stochastic process,* a fact pointed out much earlier in the context of stochastic models with either correlated or uncorrelated noise (Hänggi and Thomas, 1977). As a hypothetical example, to assert that prices are lognormal, without stating the pair correlations or transition density, does not tell us that prices are generated by an Ito process.

6.2 Selfsimilar Ito processes

This first subsection is general, is not restricted to Ito processes. Consider drift-free trajectories that scale with some function of time,

$$x(t) = \sigma_1(t)x(1) \tag{6.7}$$

Notice that $\sigma_1(0) = 0$ is necessary for scaling, so that all trajectories that scale necessarily pass through the origin. Hurst exponent scaling is defined by $\sigma_1(t) = t^H$.

With averages given by

$$\langle x^n(t) \rangle = \int x^n f_1(x,t) \, dx \tag{6.8}$$

then the variance is

$$\sigma^2(t) = \langle x^2(t) \rangle = \sigma_1^2(t) \langle x^2(1) \rangle \tag{6.9}$$

Satisfying (6.8) requires that

$$f_1(x,t) = \sigma_1^{-1}(t)F(u), \quad u = x/\sigma_1(t) \tag{6.10}$$

Next, we show that Hurst exponent scaling is the only possibility for a selfsimilar process.

First, let

$$x(t) = bx(at) \tag{6.11}$$

be a general selfsimilar process. From

$$x(t') = b(a')x(a't) \tag{6.12}$$

and $t' = a't$ we obtain

$$x(t) = b(a'a)x(a'at) = b(a')b(a)x(a'at) \tag{6.13}$$

so that

$$b(a) = a^H \tag{6.14}$$

with $H > 0$ follows. Setting $at = 1$ in (6.11) we obtain

$$x(t) = t^H x(1) \tag{6.15}$$

We can therefore take $\phi_1(t) = t^H$ for any selfsimilar process (Embrechts and Maejima, 2002). So far our conclusions are not restricted to Ito processes but apply to any selfsimilar process. Next, we consider selfsimilar diffusive processes.

6.2.1 Martingales

The sde

$$dx = \sqrt{D(x,t)}dB(t) \tag{6.16}$$

generates a drift-free Ito process $x(t)$ and transforms one-to-one with the Fokker–Planck pde

$$\frac{\partial p_2}{\partial t} = \frac{1}{2}\frac{\partial^2 (Dp_2)}{\partial x^2} \tag{6.17}$$

for the transition density of the Ito process. Scaling of the one-point density $f_1(x,t)$ combined with the sde (6.16) yields variance scaling

$$\sigma^2(t) = \int_0^t ds \int_{-\infty}^{\infty} dx f_1(x,s) D(x,s) = t^{2H} \langle x^2(1) \rangle \tag{6.18}$$

if and only if the diffusion coefficient scales as

$$D(x,t) = t^{2H-1} \bar{D}(u) \tag{6.19}$$

Scaling is restricted to the one-point density $f_1(x,t) = p_2(x,t|0,0)$, where the scale-independent part $F(u)$ satisfies the ode

$$2H(uF(u))' + (\bar{D}(u)F(u))'' = 0 \tag{6.20}$$

which is solved by

$$F(u) = \frac{C}{\bar{D}(u)} e^{-2H \int u du / \bar{D}(u)} \tag{6.21}$$

6.2 Selfsimilar Ito processes

if no current J flows through the system, otherwise there's another term proportional to J. We can easily calculate some examples.

First, let $\bar{D}$= constant. Then

$$F(u) = \left(\frac{H}{\bar{D}\pi}\right)^{1/2} e^{-Hu^2/\bar{D}} = \left(\frac{1}{2\pi\langle x^2(1)\rangle}\right)^{1/2} e^{-u^2/2\langle x^2(1)\rangle} \tag{6.22}$$

Second, assume that

$$\bar{D}(u) = 1 + |u| \tag{6.23}$$

Here, we find that

$$F(u) = \frac{C}{\bar{D}(u)^{1-2H}} e^{-2H|u|} \tag{6.24}$$

Both these results are changed if we include a constant factor multiplying the diffusion coefficient, but we leave the details as an exercise. Here, we obtain a pure exponential density if and only if $H = 1/2$, otherwise there is a u-dependent prefactor. In any case there are no fat tails; the exponential factor dominates as u increases in magnitude.

To obtain

$$F(u) = Ce^{-|u|} \tag{6.25}$$

we need

$$\widehat{D}(u) = \frac{1}{2H}(1 + |u|) \tag{6.26}$$

Last, we study the class of quadratic diffusion coefficients

$$\bar{D}(u) = d'(\varepsilon)(1 + \varepsilon u^2) \tag{6.27}$$

which yields the two-parameter (ε, H) class of student-t-like densities

$$F(u) = C'(1 + \varepsilon u^2)^{-1-H/\varepsilon d'(\varepsilon)} \tag{6.28}$$

where H and ε are independent parameters to be determined empirically. Here we have fat tails,

$$F(u) \approx |u|^{-\mu}, \ |u| \gg 1 \tag{6.29}$$

with tail exponent $\mu = 2 + 2H/\varepsilon d'(\varepsilon)$. We can generate all fat tail exponents in the range $2 < \mu < \infty$, but the variance is finite, $\sigma^2 = ct^{2H}$, if and only if $\mu > 3$. For $2 \leq \mu \leq 3$ the variance is infinite.

The original references on scaling in diffusive processsses are Alejandro-Quinones *et al.* (2005, 2006), Bassler *et al.* (2006), and Gunaratne and McCauley (2005a).

Finally, a variable drift $R(x,t)$ can be invented to satisfy scaling but the drift so constructed violates the variable drift of Martingale option-pricing theory. The Martingales above can be transformed into Ito processes with t-dependent drift $R(t)$ by the substitution $x \to x - \int R(t)dt$.

6.2.2 What does H ≠ 1/2 mean?

The above solutions are Martingales. We know that the increments of Martingales are nonstationary when the variance is nonlinear in t. For a scaling process this requires $H \neq 1/2$. Vanishing increment autocorrelations is easy to prove using the Martingale increment formula

$$x(t,T) = \int_t^{t+T} \sqrt{D(x(s),s)}dB(s) \tag{6.30}$$

There's no autocorrelation in the increments $x(t,-T)$, $x(t,T)$ over the two nonoverlapping time intervals,

$$\langle x(t,-T)x(t,T) \rangle = \int_t^{t+T} ds \int_{t-T}^{t} dw \langle D(x(w),w)D(x(s),s) \rangle \langle dB(w)dB(s) \rangle = 0 \tag{6.31}$$

because $\langle dB(w)dB(s) \rangle = 0$ for nonoverlapping time intervals dw and ds. The function $D(x,t)$ is called "nonanticipating" in the math literature. This just means that, by Ito's definition of the stochastic integral (6.30), the function $D(x,t)$ of the random variable x and the random increment $dB(t)$ from t to $t + dt$ are statistically dependent because $x(t)$ was determined in the Martingale sde by the Wiener increment $dB(t - dt)$ before $dB(t)$ occurs. That is, $D(x(t),t)$ cannot "anticipate" the next random increment $dB(t)$ in (6.30).

It's also easy to illustrate the nonstationarity of the increments when $H \neq 1/2$ and for $H = 1/2$ as well:

$$x(t+T) - x(t) = \int_0^{t+T} \sqrt{D(x(s),s)}dB(s) - \int_0^{t} \sqrt{D(x(s),s)}dB(s)$$

$$= \int_t^{t+T} \sqrt{D(x(s),s)}dB(s) = \int_t^{t+T} |s|^{H-1/2}\sqrt{D(u)}dB(s) \tag{6.32}$$

$$= \int_0^{T} |s+t|^{H-1/2}\sqrt{D(x/|s+t|^H)}dB(s) \neq x(T)$$

where $B(t)$ has stationary increments, $dB(s+t) = dB(s)$.

Scaling Martingales with nonstationary increments have an "intrinsic time," $\tau = t^{2H}$. If we transform a scaling Martingale to this time variable then the variance is linear in τ and the mean square fluctuation in the increment becomes stationary even if (6.32) generally is not.

So $H \neq 1/2$ simply means nonstationary increments if the increments are uncorrelated.

With $H = 1/2$ the variance is linear in the time t. This is a necessary but insufficient condition for stationary increments. Stationary increments are not proven for Martingales scaling with $H = 1/2$, and cannot be assumed to hold.

Next, we'll show that the implication of $H \neq 1/2$ is quite different for processes with stationary correlated increments.

6.3 Long time increment correlations

We consider next the class of stochastic processes generating long-time increment autocorrelations, both without and with scaling.

Let $x(t,T) = x(t+T) - x(t)$ denote an increment and let $x(t,-T)$ denote the preceding increment. We obtain

$$2\langle x(t,-T)x(t,T)\rangle = \langle (x(t,-T)+x(t,T))^2\rangle - \langle x^2(t,T)\rangle - \langle x^2(t,T)\rangle \quad (6.33)$$

If the stochastic process $x(t)$ has *stationary increments*, $x(t,T) = x(t+T) - x(t) = x(0,T)$, meaning that the one-point distribution of $x(t,T)$ is independent of t. Then (using the sloppy but standard notation $x(0,T) = x(T)$) the mean square fluctuation calculated from *any* starting point $x(t)$ is independent of starting time t,

$$\langle (x(t+T) - x(t))^2 \rangle = \langle x^2(T) \rangle \quad (6.34)$$

From this result we obtain long time increment autocorrelations

$$2\langle x(t,-T)x(t,T)\rangle = \langle x^2(2T)\rangle - 2\langle x^2(T)\rangle \quad (6.35)$$

if and only if the variance is nonlinear in the time. Exactly *how* the variance is nonlinear in t is irrelevant: the sole ingredients required for long time increment autocorrelations are:

(i) stationary increments, $x(t,T) = x(T)$ "in distribution"
(ii) a variance nonlinear in the time.

If, *in addition*, the variance would scale in time with H, *then* we would obtain the simple prediction

$$\left\langle (x(t+T) - x(t))^2 \right\rangle = \left\langle x^2(T) \right\rangle = cT^{2H} \tag{6.36}$$

and also that

$$2\langle x(t,-T)x(t,T)\rangle / \langle x^2(T)\rangle = 2^{2H-1} - 1 \tag{6.37}$$

Clearly, scaling with arbitrary Hurst exponent $0 < H < 1$ is not a condition for the presence or absence of long time correlations. Next, we present the canonical example where both long time pair correlations and scaling appear simultaneously. The main references here are McCauley *et al.* (2007a, 2007c), who followed Mandelbrot and van Ness (1968).

6.3.1 Fractional Brownian motion defined as a stochastic integral

For finite autocorrelations over nonoverlapping time intervals the increments must be stationary,

$$x(t+T) - x(t) = x(T) \tag{6.38}$$

so that, for example,

$$\langle (x(t+T) - x(t))^n \rangle = \langle x(T)^n \rangle \tag{6.39}$$

for $n = 1, 2, 3, \ldots$. To try to construct such a process, consider stochastic integrals of the form

$$x(t) = \int_{t_0}^{t} k(t,s) \mathrm{d}B(s) \tag{6.40}$$

With $k(t,s)$ dependent on the upper limit t, the stochastic integrals cannot be generated by an sde. This condition avoids the construction of an Ito process. We know from the discussion above that *if* the increments of (6.40) are stationary, *then* long time autocorrelations in increments appear if the variance is nonlinear in t. Stationary increments apparently occur if and only if $t_0 = -\infty$ in (6.40). Satisfying the stationarity condition

$$x(t+T) - x(t) = \int_{-\infty}^{T} k(T,s) \mathrm{d}B(s) = x(T) \tag{6.41}$$

requires a very special class of kernels $k(t,s)$. We have not assumed that $x(t)$ scales.

Independently of the question of stationarity of the increments and the corresponding long time autocorrelations, consider next the possibility of scaling, $x(t) = t^H x(1)$. Transforming variables $u = s/t$ in (6.40) we obtain

$$x(t) = \int_{t_0/t}^{1} k(t, tu) t^{1/2} dB(u) \tag{6.42}$$

because $dB(s) = B(s + ds) - B(s) = B(ds)$ so that $B(tdu) = t^{1/2}B(du) = t^{1/2}dB(u)$. To get $x(t) = t^H x(1)$ we need both that the kernel scales, $k(t, tu) = t^{H-1/2} k(1, u)$, and that the lower limit of integration is either $t_0 = 0$ or $-\infty$. For the former case the increments of (6.42) are typically not stationary, but one may obtain stationary increments for $t_0 = -\infty$ for a very special kernel $k(t,s)$. In either case, with or without long time autocorrelations, we have a stochastic process that scales.

The main point here is that in order to obtain the standard predictions for the long time correlations of fBm where

$$\langle (x(t+T) - x(t))^n \rangle = \langle x(T)^n \rangle = cT^{nH} \tag{6.43}$$

two entirely separate conditions must be satisfied. First, the increments must be stationary, for without this condition scaling merely leads to Markov-like behavior. Second, the variance must scale with H. These two separate conditions are generally confused together in the literature with far too much emphasis on the second one. To test either of these two assumptions empirically correctly is more difficult than most of the existing literature on the subject would have the reader believe. The key test is for stationarity and correlation of the increments, not scaling.

However, if we have stationary increments combined with Hurst exponent scaling, then a simple prediction for the autocorrelations of fBm over non-overlapping time intervals follows easily. Let $t_1 - T_1 < t_1 < t_2 < t_2 + T_2$. With the autocorrelation function defined by

$$\begin{aligned} 2\langle (x(t_2 + T_2) - x(t_2))(x(t_1) - x(t_1 - T_1)) \rangle \\ = \langle (x(t_2 + T_2) - x(t_1 - T_1))^2 \rangle + \langle (x(t_2) - x(t_1))^2 \rangle \\ - \langle (x(t_2 + T_2) - x(t_1))^2 \rangle - \langle (x(t_2) - x(t_1 - T_1))^2 \rangle \end{aligned} \tag{6.44}$$

where we've used $2(a - c)(d - b) = (a - b)^2 + (c - d)^2 - (a - d)^2 - (c - b)^2$, then using stationarity of the increments, and also dividing by a product of the variances at times T_1 and T_2 with $t_2 = t/2 = -t_1$, we can evaluate

$$\begin{aligned} 2C(S_1, S_2) = \langle (x(t/2 + T_2) - x(t/2))(x(-t/2) - x(-t/2 - T_1)) \rangle \\ / (\langle x^2(T_1) \rangle \langle x^2(T_2) \rangle)^{1/2} \end{aligned} \tag{6.45}$$

where $S_1 = T_1/t$, $S_2 = T_2/t$, to obtain

$$C(S_1, S_2) = [(1+S_1+S_2)^{2H} + 1 - (1+S_1)^{2H} - (1+S_2)^{2H}]/2(S_1 S_2)^H \quad (6.46)$$

This result was first derived by Mandelbrot and van Ness (1968). The resulting long time correlations vanish if and only if $H = 1/2$, if the variance is linear in t.

Mandelbrot and van Ness have also provided us with an example of a scaling kernel that generates stationary increments, and hence describes fBm,

$$x_H(t) = \int_{-\infty}^{0} [(t-s)^{H-1/2} - (-s)^{H-1/2}]dB(s) + \int_{0}^{t} [(t-s)^{H-1/2}dB(s) \quad (6.47)$$

or

$$x_H(t) = \int_{-\infty}^{t} [(t-s)^{H-1/2} - N(s)(-s)^{H-1/2}]dB(s) \quad (6.48)$$

where $N(s) = 1 - \theta(s)$. To see that the increments are indeed stationary, use $u = s - t$ and $s = u - t$ respectively in

$$x_H(t+T) - x_H(t) = \int_{-\infty}^{t+T} [(t+T-s)^{H-1/2} - N(s)(-s)^{H-1/2}]dB(s) \\ - \int_{-\infty}^{t} [(t-s)^{H-1/2} - N(s)(-s)^{H-1/2}]dB(s) \quad (6.49)$$

along with $dB(t+u) = B(t+u+du) - B(t+u) = B(du) = dB(u)$ to obtain

$$x_H(t+T) - x_H(t) = \int_{-\infty}^{T} [(T-u)^{H-1/2} - N(u)(-u)^{H-1/2}]dB(u) = x_H(T) \quad (6.50)$$

This result follows from a cancellation of terms from each integral in (6.49). Again, the reader should be aware that by "$x(T)$" we always mean the increment $x(0,T)$; "$x(T)$" is always measured from the origin $x(0) = 0$.

6.3.2 The distribution of fractional Brownian motion

It's easy and instructive to construct the one-point density that describes fBm,

$$f_1(x,t) = \langle \delta(x - k \bullet \Delta B) \rangle = \frac{1}{2\pi} \int_{-\infty}^{\infty} e^{ipx} \langle e^{-ipk \bullet \Delta B} \rangle dp \quad (6.51)$$

where $k \bullet \Delta B$ denotes the Ito product representing the stochastic integral (6.48). From this one easily obtains a scaling Gaussian $f(x,t) = t^{-H}F(u)$, $F(u) = (1/(2\pi \langle x^2(1) \rangle))^{1/2} \exp(-u^2/2\langle x^2(1) \rangle)$ with $u = x/t^H$, which is identical with the one-point density of a scaling Gaussian Markov process, or for a Gaussian

Martingale with finite memory. A scaling Ito process cannot be distinguished from fBm on the basis of the one-point density; the one-point density provides us with no knowledge of the underlying dynamics of the process that generated it. It's necessary to ask if the increments are stationary or uncorrelated when $H \neq 1/2$, and answering that question demands constructing the two-point density, or at least the pair correlations.

We can construct the two-point density that defines fBm because, for a Gaussian process, the pair correlations specify all densities of all orders n. FBm is a Gaussian process simply because there is no x-dependence in the kernel of the stochastic integral for fBm. However, fBm is not merely a time-transformation on a Wiener process; that triviality is eliminated by the fact that the kernel k depends on the upper limit of integration.

Any two-point Gaussian density is given by

$$f_2(x,t) = \frac{1}{2\pi \det B} e^{-x^\dagger B^{-1} x} \tag{6.52}$$

where

$$B_{kl} = \langle x_k x_l \rangle \tag{6.53}$$

defines the autocorrelation matrix. Without specifying the autocorrelations (6.53), one cannot say whether the process $x(t)$ is Markovian or not. The autocorrelation

$$\langle x(s)x(t) \rangle = \frac{\langle x^2(1) \rangle}{2} (|s|^{2H} + |t|^{2H} - |s-t|^{2H}) \tag{6.54}$$

enforces stationary increments, where scaling with H is also asssumed in agreement with (6.47), and therefore will enforce the long time autocorrelations of fBm in the increments. The resulting two-point density of fBm can be written as[1]

$$f_2(x(s),s;x(t),t) = \frac{1}{2\pi\sigma_1\sigma_2(1-\rho^2)^{1/2}} e^{-(x^2(s)/\sigma_1^2 + x^2(t)/\sigma_2^2 - 2\rho x(s)xs(t)/\sigma_1\sigma_2)/2(1-\rho^2)^2} \tag{6.55}$$

where $\sigma_1\sigma_2\rho = \langle x(s)x(t)\rangle$, $\sigma_1^2 = \langle x(1)\rangle|t|^{2H}Z$, and $\sigma_2^2 = \langle x(1)\rangle|s|^{2H}$, and

$$\rho = (|s|^{2H} + |t|^{2H} - |t-s|^{2H})/2|st|^H \tag{6.56}$$

If we integrate over the earlier variable $x(s)$, taking $s < t$, then we obtain the one-point density f_1,

---

[1] This corrects a misstatement about fBm in McCauley et al. (2007a).

$$f_1(x,t,-\infty) = \frac{t^{-H}}{\sqrt{2\pi\langle x^2(1)\rangle}} e^{-x^2/2\langle x^2(1)\rangle t^{2H}} \tag{6.57}$$

which scales with H. The result is identical with the density for a scaling diffusive process. However, the transition density $p_2 = f_2/f_1$ of fBm does *not* satisfy a diffusion pde (the reader is invited to construct the transition density).

For the analysis of time series, the two central questions are those of nonstationary vs stationary increments, and correlated vs uncorrelated increments. Scaling makes modeling easier but can't be counted on to exist in empirical data. We emphasize that (i) a Hurst exponent H, taken alone, tells us nothing about the dynamics, and even worse (ii) a one-point density, taken alone, tells us little or nothing about the dynamics. It's absolutely necessary to study the autocorrelations of increments in order to obtain any idea what sort of dynamics are generated by a time series.

Finally, we can now easily write down the two-point transition density

$$p_2(y,s|x,t) = f_2(y,s;x,t)/f_1(x,t) \tag{6.58}$$

and it to show that the conditional expectation of y is not x, but is rather

$$\int_{-\infty}^{\infty} p_2(y,s|x,t)y\,dy = C(s,t)x \tag{6.59}$$

where $C \neq 1$ varies, fBm is not a Martingale. Depending on H, the prefactor C may be either positive or negative. The factor $C(t,s)$ is proportional to the autocorrelation function that reflects the stationary increments of fBm. With $\langle x(s)\rangle = C(s,t)x$, where x is the last observed point at time t, we therefore obtain $d\langle x(s)\rangle/ds = xdC(s,t)/ds \neq 0$, showing a trend/bias that is inherent in the process and so can't be eliminated by subtracting a drift term.

FBm has infinite memory (McCauley, 2008b); the entire past trajectory is remembered in (6.48). Another way to state this is that the n-point density hierarchy f_n doesn't truncate for any finite integer n. Next, we consider processes where memory of a *finite* number of states is possible. These are Ito processes, processes consisting of a drift A plus a Martingale M; that is, $x(t) = A(t) + M(t)$. Markov processes are Ito processes with memory only of the last observed state, and of no earlier states: $R(x,t)$ and $D(x,t)$ cannot depend on any states other than (x,t). The dependence of either on a functional like $\langle x(t)\rangle = \int dx x f_1(x,t)$ is also merely state-dependence (McCauley, 2008b). A Martingale has uncorrelated, generally nonstationary increments. Adding drift to a Martingale can create increment autocorrelations.

6.4 The minimal description of dynamics

A one-point density doesn't classify the underlying dynamics (Hänggi et al., 1978). Given a one-point density, or a diffusive pde for a one-point density, we cannot even conclude that we have a diffusive process at the level of the transition density (McCauley et al., 2007a). The one-point density for fBm, a non-diffusive process with long time increment autocorrelations, satisfies the same diffusive pde as does a Gaussian Markov process. A detrended diffusive process has no increment autocorrelations, so that the pde for the transition density is also diffusive (Fokker–Planck). Therefore, the minimal knowledge needed to identify the class of dynamics is either the transition density depending on two points or else the specification of the pair correlations $\langle x(s)x(t) \rangle$.

For a general stochastic process, transition densities depending on the history of all possible lengths are required. The pair correlations are adequate to pin down the stochastic process in exactly two distinct cases. First, for a drift-free process, if $\langle x(s)x(t) \rangle = \langle x^2(s) \rangle$ with $s < t$, then the process is either Markovian (memory-free) or else is a Martingale with finite memory. In either case the process is diffusive. The other case where pair correlations determine the process is in the case of Gaussian processes. There, pair correlations specify processes of all orders.

6.5 Scaling of correlations and conditional probabilities?

Simple averages scale if the one-point density scales, and vice versa. But scaling starts and stops with one-point densities; neither transition densities nor pair correlations may scale except in one pathological case (Bassler et al., 2008).

Conditional averages and pair correlations require the two-point density

$$f_2(x, t+T; y, t) = p_2(x, t+T|y, t)f_1(y, t) \qquad (6.60)$$

or, more to the point, the two-point transition density (conditional probability density) $p_2(x, t+T|y, t)$. Assume in all that follows that $f_1(x,t) = |t|^{-H}F(x/|t|^H)$, but assume nothing in advance about the underlying dynamics.

Without specifying the dynamics, the vanishing of an unconditioned average of x does not mean that there's no conditional trend: in fBm, for example, where $\langle x(t) \rangle = x(0) = 0$ by construction, the conditional average of x does not vanish and depends on t, reflecting either a trend or an anti-trend. In a Martingale process, scaling may occur if the drift rate is either constant or depends on t alone (is independent of x) and has been subtracted, that by "x" we really mean the detrended variable $x(t) - \int R(s) ds$. Markov processes with x-independent drift can be detrended over a definite time scale, but any attempt to detrend fBm would at best produce spurious results because the "trend" is due to long time autocorrelations, not to a removable additive drift term.

Here's the main point of this section. Even if scaling holds at the one-point level as in fBm, then the two-point density (the transition density p_2) and the pair correlations $\langle x(t)x(s)\rangle$

$$\langle x(t)x(s)\rangle = \int\int dydx\, yx f_2(y,t;x,s) \tag{6.61}$$

do *not* scale with H in the two different times (t,s), and it's the transition density p_2, or at least the pair correlations, that's required to give a minimal description of the underlying dynamics.[2] We can illustrate this via some closed-form examples.

We know how to calculate f_2 and p_2 analytically only for Gaussian processes, where the densities of all orders are determined once the pair correlations are specified, so let's examine Hurst exponent scaling in that case. Assume that the process is both Gaussian,

$$p_2(x,t|y,s) = \frac{1}{\sqrt{2\pi K(t,s)}} e^{-(x-m(t,s)y)^2/2K(t,s)} \tag{6.62}$$

$$f_1(y,s) = \frac{1}{\sqrt{2\pi\sigma^2(s)}} e^{-x^2/2\sigma^2(s)} \tag{6.63}$$

and selfsimilar. The selfsimilarity condition requires $\sigma^2(t) = t^{2H}\langle x^2(1)\rangle$. This immediately yields scaling of f_1, $f_1(y,s) = |s|^{-H} F(y/|s|^H)$, but nothing else.

Consider next three separate cases. First, assume statistical independence of the process at different times (t,s), $f_2(x,t;y,s) = f_1(x,t)f_1(y,s)$, so that if f_1 scales then so does f_2, $f_2(x,t;y,s) = t^{-H}F(x/t^H)s^{-H}F(y/s^H)$, but $m(t,s) = 0$, $\langle x(t)x(s)\rangle = 0$. This is the trivial case.

Next, because

$$\langle x(t)\rangle_{cond} = \int_{-\infty}^{\infty} dx\, x p_2(x,t|y,s) = m(t,s)y \tag{6.64}$$

for Martingale dynamics we must require $m(t,s) = 1$ and so we obtain the pair correlations $\langle x(t)x(s)\rangle = \langle x^2(s)\rangle = |s|^{2H}\langle x^2(1)\rangle$, $t > s$. For a Gaussian Martingale neither the pair corelations nor p_2 scales in both t and s.

---

[2] For a Gaussian process, pair correlations and p_2 provide a complete description. But for non-Gaussian processes like FX markets, all of the transition densities p_n, $n = 2,3,\ldots$ may be required to pin down the dynamics. In data analysis it's hard to measure more than pair correlations.

6.5 Scaling of correlations and conditional probabilities?

Next, apply

$$\langle x(t)x(s)\rangle = m(t,s)\sigma^2(s) \tag{6.65}$$

to fBm, a selfsimilar Gaussian process with stationary increments. Here,

$$\langle x(t)x(s)\rangle = \frac{\langle x^2(1)\rangle}{2}(|s|^{2H}+|t|^{2H}-|s-t|^{2H}) \tag{6.66}$$

follows. This is the canonical "selfsimilar process with long time autocorrelations," but neither the pair correlations (6.65), the increment autocorrelations, nor the transition density (6.62) scale in both times t and s.

From (6.62) and (6.65) we see that, excepting the trivial case of statistical independence where the pair correlations vanish, scaling of both the pair correlations and f_2 occurs if and only if the pathology $m(t,s) = |t|^H/|s|^H$ holds. *We therefore conjecture from these Gaussian examples that, in general, neither pair correlations nor two-point (or higher order) densities scale whenever the stochastic process is selfsimilar.* This means that scaling does not have the importance for understanding market dynamics that econophysicists have heretofore assumed. In particular, when long time correlations are claimed in the literature then the reader should ask: what is the evidence presented for stationary increments? In the next chapter we'll discuss the questions of increment autocorrelations and increment stationarity in finance data.

7
Statistical ensembles
Deducing dynamics from time series

We now begin our ascent toward the main peak of our effort, although there are still interesting and useful peaks to climb in the remaining chapters. The theories of stochastic processes and probability are put to work below to address the central question: can we reliably deduce a model or class of dynamic models from a single time series, where "systematically rerunning the experiment" is impossible? If so, then how, and what are the main pitfalls to be avoided along the way? With detrended data in mind, the two classes of dynamics of interest are those with and without increment autocorrelations: Martingales vs everything else. We will also see that Wigner's analysis applies (Chapter 1): unless we can find an inherent statistical repetitiveness to exploit, then the effort is doomed in advance. We will exhibit the required statistical repetition for FX data, and also show how a class of diffusive models is implied. Because we work with detrended time series (this would be impossible were the increments correlated), attention must first be paid to restrictions on detrending Ito processes.

7.1 Detrending economic variables

Prices are recorded directly but we'll study log returns of prices. The use of logarithms of prices is common both in finance and macroeconomics. Before the transformation can be made from price to log returns, a price scale must be defined so that the argument of the logarithm is dimensionless. In finance, we can begin with the sde for price p, $dp = \mu p dt + p\sqrt{d(p,t)}dB$, where p can be detrended multiplicatively so that $S = pe^{-\mu t}$ is a Martingale. If we next transform to log returns, then we could arbitrarily define $x = \ln(p(t)/V(t))$ where $V(t)$ is nonunique, but here we choose the price $V(t)$ to locate the minimum of the price diffusion coefficient $e(S,t)$ in $dS = S\sqrt{e(S,t)}dB$, so that $V(t) = Ve^{-\mu t}$ where V would locate that minimum at $t = 0$. The corresponding Fokker–Planck pde for the log returns Green function is then

$$\frac{\partial g}{\partial t} = \frac{1}{2}\frac{\partial}{\partial x}(D(x,t)g) + \frac{1}{2}\frac{\partial^2}{\partial x^2}(D(x,t)g) \tag{7.1}$$

and we get corresponding log increments $x(t, T) = \ln(e^{-\mu T} p(t+T)/p(t))$. *If* there is but a single price scale in the problem, *then* V is the consensus price at $t = 0$, where the consensus price locates the peak of the one-point density.

In FX data analysis, the drift term is so small that it nearly can be ignored. Because of that, in our FX data analysis below we can effectively work with the approximation $x(t, T) \approx \ln(p(t+T)/p(t))$. In this approximation the returns were detrended empirically, corresponding to an approximate Martingale Fokker–Planck pde in returns *also* with the variable drift term in (7.1) ignored. This completely drift-free approximation was used in our FX data analysis (Bassler *et al.*, 2007) described below. Detrending of prices and returns is discussed formally in McCauley *et al.* (2007c).

In practice, detrending a price series directly is problematic unless the time scale is small enough that we can linearize the exponential multiplying $p(t)$ and then simply subtract the drift term from the price increments. The reason for this is that we don't know the unreliable parameter µ in advance, and generally have no good way to determine it. In our discussions of increments below we always assume detrended time series, because a drift term can generate trivial increment autocorrelations.

7.2 Ensemble averages constructed from time series

How and when can the dynamics that generated a single time series be discovered via a reliable statistical analysis? What constitutes a reliable statistical analysis? Which quantities should be measured, and which should be ignored? The standard methods of statistics and econometrics fail to shed light on market dynamics, and may even generate spurious results. How can we do better? We'll pose and answer the main question for the social sciences in particular, and for modeling in general: given a single historical record in the form of a price series, we can construct ensemble averages and discover evidence of a definite underlying law of motion if statistical regularity can be discovered in the time series. If statistical regularity cannot be found then a reliable statistical analysis of a single, historic time series may not be possible. We'll see that the analysis of a single, historic time series is nontrivial, and that a statistical ensemble can be constructed if there is an underlying time scale for treating a collection of equal time segments of the series approximately as "reruns of one uncontrolled experiment."

To demonstrate these assertions, we begin by explaining why the construction of a statistical ensemble is a necessary condition for time series analysis in the first place. Toward that end, we begin with the requirements imposed by limited, finite precision in measurement, namely, the binning of data on the real axis, the construction of histograms, and statistical averages.

The original references for this chapter are Bassler *et al.* (2007, 2008) and McCauley (2008c).

7.2.1 Coarsegraining

Consider the time series generated by a one-dimensional stochastic process. Coarsegrain the *x*-axis into bins. The number and size of the bins must be such that, excepting the region for large enough *x* where few or no points are observed, the number of points per bin is large compared with unity. Obviously, this will fail when *x* is large enough in magnitude: "good statistics" means having many points in each bin. As Feynman wrote, there's a very good reason why the last measured point is not reliable.

For good histograms ("good statistics") we need many points at *each* time *t*. One therefore needs many reruns of the same, identical experiment in order to obtain good statistical averages. We need N different realizations of the process $x_k(t)$, $k = 1, \ldots, N$, where for good statistics $N \gg 1$. At time *t* each point in each run adds one point to the histogram. The average of a dynamical variable $A(x,t)$ is then given by

$$\langle A(x,t) \rangle = \frac{1}{N} \sum_{k=1}^{N} A(x_k(t), t) \tag{7.2}$$

where the N values $x_k(t)$ are taken from different runs repeated *at the same time t*, resetting the clock to $t = 0$ after each run.

Assume that the variable *x* takes on *n* discrete values x_m, $m = 1, 2, \ldots, n$, and assume that x_m occurs W_m times during the N runs and falls into the *m*th bin, and denote $w_m = W_m/N$, $\sum_{m=1}^{n} W_m = 1$, with

$$N = \sum_{m=1}^{n} W_m \tag{7.3}$$

Then

$$\langle A(x,t) \rangle = \sum_{m=1}^{n} w_m A(x_m(t), t) \tag{7.4}$$

The w_m are what we mean by the histograms for the one-point density. If the histograms can be approximated by a smooth density $f_1(x,t)$, then (7.4) becomes

7.2 Ensemble averages constructed from time series

$$\langle A(x,t) \rangle = \int dx f_1(x,t) A(x,t) \tag{7.5}$$

as n goes to infinity. This is the unconditioned ensemble average. In general, an ensemble average is an average over a density at one time t, and is generally different than a time average. In reality N and n are finite. An empirical average (7.4) should show scatter about the ensemble average (7.5) as t is varied (such scatter is shown in Figure 7.2 below).

We always adhere to what Gnedenko (1967) calls "the statistical definition of probability," the definition in terms of relative frequencies observed in a large number of repetitions of the same experiment. Stratonovich (1963) has stated that, given $x_1,\ldots,x_N$, the realizations of a random variable (e.g. $x(t)$ for fixed t), the mean can be defined by

$$\langle x \rangle = \frac{x_1 + \ldots + x_N}{N} \tag{7.6}$$

as the arithmetic mean of the sample values as the number of sample values is increased without limit. Probability theory can only be used to study experimental data for which such limits exist and do not depend on how the realizations $x_1,\ldots,x_N$ are chosen from the total statistical ensemble. We want next to make this precise, to explain why ensemble averages make sense in light of Tschebychev's Theorem.

7.2.2 Statistical ensembles

We begin with the case where laboratory experiments are possible, because this viewpoint provides the statistical theory that must be approximated in economics and finance. The repeated runs of an experiment would allow us to define "statistical probability" and the corresponding averages, and Tschebychev's Theorem (Chapter 3) will be used to make the idea precise. In what follows, the process may be nonstationary with nonstationary increments; there is no restriction to any kind of stationarity.

Let there be N experimental realizations of a time series $x(t)$, where the system is strobed at the same times $t_1, t_2, \ldots, t_N$ in each run. This is possible when studying turbulence in a wind tunnel, for example, but not in astronomy and economics, nor is it possible when studying atmospheric turbulence or the hydrodynamics of the ocean. Consider the N points $x_k = x_k(t)$, $k = 1, \ldots, N$, for the runs at the same time t. Then the histogram for the one-point density at one time t is given by

$$f_1(x,t) \approx \frac{1}{N} \sum_{k=1}^{N} \delta(x - x_k) \tag{7.7}$$

and will show scatter so long as N is finite, which is necessarily the case in experiments and simulations. To apply Tschebychev's Theorem for convergence ("convergence" in practice means systematically reduced scatter as N is increased), we need

$$\langle \delta(x - x(t))\delta(y - x(t))\rangle = 0 \tag{7.8}$$

The ensemble average is

$$\langle \delta(x - x(t))\delta(y - x(t))\rangle = \iint dx_1 dx_2 \delta(x - x_1)\delta(y - x_2) f_2(x_2, t; x_1, t) \tag{7.9}$$

With $f_2(y, t; x, s) = p_2(y, t|x, s) f_1(x, s)$ and $p_2(y, t|x, t) = \delta(y - x)$ we obtain vanishing correlations (7.8) for $y \neq x$, so the fixed-t series (7.7) for the histograms converges to the ensemble average $f_1(x,t)$ as N increases. One can show similarly that the correlations of other quantities calculated at equal times vanish as well. *Tschebychev's Theorem provides the basis for the construction of statistical ensembles for general stationary and nonstationary processes.* The interesting question for us is: how can we implement the idea of ensemble averages when we're faced with a single, historic time series?

7.3 Time series analysis

In finance and economics we cannot "rerun the experiment." We have but a single, historic time series for each different macroeconomic variable. A single time series provides us *a priori* with no statistics, no histograms: there is only one point at each time t. This does not present a big difficulty in astronomy and atmospheric turbulence because we know the laws of motion. In economics and finance, we do not know *anything* in advance (*a priori* expectations must be distinguished from knowledge); discovering the correct underlying dynamics is *the* problem to be solved (this is not the viewpoint of econometrics!). The only option available to us is that we meet time series that can be understood approximately as N statistically equivalent pieces, where a time scale for some sort of statistical regularity can be found. If the increments would be stationary, then N could be taken to be arbitrary, but stationary increments cannot be established without an ensemble analysis. Also, the increments cannot be expected or assumed to be stationary; most time series should be expected to exhibit nonstationary increments.

For an ergodic stationary process the problem could very easily be solved. There, a single time series can be used to obtain statistics reflecting an entire ensemble (Gnedenko, 1967). For the case of nonstationary processes, however, there is no ergodic theorem. Nonstationary processes can be classified as those having stationary increments, and those that do not. We now describe the analyses required in all of these cases, beginning with the easiest one.

7.3.1 Ergodic stationary processes

If the underlying process is time-translationally invariant, then a finite length time series can be broken up into N pieces, and the size of N will not matter, although $N \gg 1$ will be needed to get good statistics. Each segment can be regarded as a rerun of the experiment from $t = 0$ to T, where NT is the total length of the original series. These pieces can then be used to calculate the ensemble average of a t-independent function $A(x)$ at each time t, $0 \leq t \leq T$. If the process is stationary, then this average will not vary with t. Instead, it will appear as a flat line (time averages for a stationary process converge to *some* limit). If, in addition, the condition for ergodicity is satisfied, then the time average converges to the ensemble average

$$\langle A \rangle = \int dx A(x) f_1(x) \tag{7.10}$$

within scatter due to the fact that N and n are finite. If the line is not flat, then the series is not stationary. Note that fat tails in f_1 can cause some variables $A(x)$ to blow up, in which case one must restrict the test to low moments. For stationary processes, time averages converge to some limit, not necessarily to the limit given by the ensemble average.

If the time series is both ergodic and stationary (Yaglom and Yaglom, 1962; Stratonovich, 1963; Gnedenko, 1967), then

$$\langle A(x) \rangle = \int dx A(x) f_1(x) \tag{7.11}$$

holds with probability one as N goes to infinity. This is a generalization of the law of large numbers, and means that for finite N the difference between the time average

$$\langle A(x) \rangle_t = \frac{1}{N} \sum_{k=1}^{N} A(x_k(t)) \tag{7.12}$$

and the ensemble average $\langle A(x) \rangle = \int dx A(x) f_1(x)$ should look like scatter. *In this case, a time average taken over a finite length series provides an estimate of the ensemble average.* This is what's meant in the literature by the statement: if a system is ergodic, then a single time series can be used to deduce the underlying statistical properties.

Tschebychev's Theorem requires that *N pairwise uncorrelated* random variables x_k have a common mean and bounded variances in order that ergodicity in the form

$$\langle x(t) \rangle = \frac{1}{N} \sum_{k=1}^{N} x_k(t) = \int \mathrm{d}x x f_1(x) \qquad (7.13)$$

holds in probability as N goes to infinity. This can be applied to functions of x so long as the functions are uncorrelated. The more general basis for the convergence of time averages of a stationary process is the Birkhoff–Khinchin–Kolmogorov ergodic theorem (Gnedenko, 1967) and can be understood as follows. With a single stationary time series divided into N equal segments, we can calculate the empirical ensemble average, which in principle should converge as N goes to infinity. The scatter in the empirically constructed ensemble average should be reduced systematically as N is increased. Consider

$$\langle x \rangle_t = \frac{1}{T} \int_0^T x(s) \mathrm{d}s \qquad (7.14)$$

Form the variance in $\langle x \rangle_t$ as an ensemble average. This yields

$$\left\langle (\langle x \rangle_t - \langle x \rangle)^2 \right\rangle_t = \frac{1}{T^2} \int_0^T \int_0^T \mathrm{d}s \mathrm{d}t \langle (\langle x \rangle_t - \langle x \rangle)(\langle x \rangle_s - \langle x \rangle) \rangle \qquad (7.15)$$

For a stationary process the pair correlation function $\langle x(t)x(t+T) \rangle$ must depend on T alone, independent of t, so that the integrand is the pair correlation function and can depend on $t - s$ alone. Denote the pair correlation function by $R(t - s)$. By a coordinate transformation the integral can be cast into the form

$$\left\langle (\langle x \rangle_t - \langle x \rangle)^2 \right\rangle_t \approx \frac{1}{T^2} \int_0^T \mathrm{d}\tau \tau R(\tau) \leq \frac{1}{T} \int_0^T \mathrm{d}\tau R(\tau) \qquad (7.16)$$

This shows that the time average converges in probability to the ensemble average as $T \to \infty$ if the pair correlation is integrable from 0 to ∞. For discrete time random variables Gnedenko shows that the condition for ergodicity is the vanishing of $R(t - s)$ as $t - s$ goes to infinity (an example will be given in Chapter 10 from regression analysis). Kac (1949) proved that discrete stationary processes are recurrent. We cannot escape the Tschebychev condition that, asymptotically, pair correlations must vanish, and in the case of continuous time, they must vanish sufficiently fast that $R(\tau)$ is integrable over the infinite interval (an example is the OU process).

Strong stationarity means that all densities in the infinite hierarchy are both normalizable and time-translationally invariant. Obviously, strong stationarity cannot be verified empirically. At best, we might hope to get histograms for f_1, but even this is difficult. The notion of weak stationarity was introduced because, even for the one-point density f_1, the empirical histograms converge notoriously slowly in practice. Weak stationarity replaces the requirements of measuring densities by the relative ease of measuring simple averages and pair correlations. Weak stationarity means that we verify, in practice, that the mean and variance are constants, and that the pair correlation function depends on time lag T alone independent of the starting time t. It may initially seem to the reader like magic, but where densities show too much scatter certain averages can still be computed reliably. We can expect this to hold for nonstationary processes as well, and will illustrate the phenomenon for FX data, where we will see that the empirical extraction of one-point densities is impossible, even on the time scale of a day (and taking longer time scales like a week only makes the statistics worse by reducing the number of points in the ensemble).

Next, we focus on nonstationary processes. In that case there is no ergodic theorem to permit the replacement of time averages by ensemble averages. We will use Tschebychev's Theorem to show that time averages cannot be expected to converge to any definite limit at all.

7.3.2 Stationary increment processes

Stationary increments "in probability" means precisely, with $z = x(t+T) - x(t) = y - x$, that the *one-point increment density*

$$f(z, t, t+T) = \iint dy dx f_2(y, t+T; x, t) \delta(z - y + x) \quad (7.17)$$

or

$$f(z, t, t+T) = \int dx f_2(x+z, t+T; x, t) \quad (7.18)$$

is independent of t and depends on the lag time T alone. There is no requirement placed on time-translational invariance of the densities f_n, $n > 2$, and (as we showed in Chapter 3) the pair correlations of stationary increment processes generally admit no ergodicity. Stationary processes trivially generate stationary increments, but arbitrary *stationary increment processes are generally nonstationary*.

Stationarity of increments could in principle be established empirically as follows. Break up a single long time series into N pieces, where N is large enough to be able to get good statistics. If the increments are stationary, the following result will hold for all N. If the increment density can be extracted from the ensemble, then it will be independent of t. For example, calculate the increment from $t = 0$ to T. Then from $t = 1$ to $1 + T$, and so on. The problem with this test is that it's in practice impossible; densities generally cannot be reliably extracted. That is, it will be very hard or even impossible to verify stationary increments even if they would occur.

The idea of weak stationarity was introduced historically because even one-point and two-point densities cannot be obtained reliably; there's too much scatter in the histograms. Economic data are certainly too sparse to test for a density exhibiting stationary increments, so we introduce the notion of *weak increment stationarity*. In weak increment stationarity, we ask only if the mean square fluctuation is time-translationally invariant; that is if $\langle x^2(t,T)\rangle = \langle x^2(0,T)\rangle = \langle x^2(1)\rangle T$. This would guarantee that the process variance is linear in t, but does not imply that the increments are stationary. First, of great interest to macroeconomists, is there any case where ergodicity follows from increment stationarity?

We already know that there is only one nonstationary process for which an ergodic theorem applies to stationary increments, the Wiener process. The Markov condition can be written as

$$f_n(x_n, t_n; x_{n-1}, t_{n-1}; \ldots; x_1, t_1) = p_2(x_n, t_n | x_{n-1}, t_{n-1}) \\ p_2(x_{n-1}, t_{n-1} | x_{n-2}, t_{n-2}) \ldots p_2(x_2, t_2 | x_1, t_1) p_2(x_1, t_1 | 0, 0) \quad (7.19)$$

and if the process is both time- and x-translationally invariant (implying the Wiener process), then with fixed time differences $T = t_k - t_{k-1}$ we obtain an iid condition on the increments $z_n = x_n - x_{n-1}$ and $T - t_1$, where $z_1 = x_1$, as

$$f_n(x_n, t_n; x_{n-1}, t_{n-1}; \ldots; x_1, t_1) = p_2(z_n, T | 0, 0) \\ p_2(z_{n-1}, T | 0, 0) \ldots p_2(z_2, T | 0, 0) p_2(z_1, T | 0, 0) \quad (7.20)$$

This means that Tschebychev's Theorem can be used to predict that time averages converge to ensemble averages. This is the only case of stationary increments where an ergodic theorem holds. This foreshadows what we are about to establish next: stationary increments or not, an ensemble analysis must be performed. In particular, we will show next that time averages in the form of a "sliding window" cannot be expected to yield reliable statistics. We show in another section below how time averages on nonstationary processes generate spurious stylized facts.

The time average of the increment density is defined by

$$f_s(z, T) = \frac{1}{N} \sum_{t=t_1}^{t_N} \delta(z - x(t, T)) \quad (7.21)$$

where in this case the delta function should be understood as the Kronecker delta. We assume stationary increments, so that $x(t,T) = x(0,T)$ "in distribution." The time average is constructed by sliding a window in the following way: start at time t in the time series. Then read the difference $x(t,T)$ at the points t and $t + T$. Since N is the number of points in the time series, to ensure uncorrelated increments we must restrict to $t = nT$. Even this restriction doesn't save the procedure from defects: the definition of the two-point (ensemble average) increment density is

$$f(z_1, t_1, t_1 + T; z_2, t_2, t_2 + T) = \langle \delta(z_1 - x(t_1, T))\delta(z_2 - x(t_2, T))\rangle \quad (7.22)$$

Two points are noteworthy. First, (i) the two-point density defined by (7.22) is not necessarily time-translationally invariant, and (ii) this density generally doesn't vanish, or even factor into two one-point increment densities for nonoverlapping time intervals. *The variables in the time series (7.21) are strongly correlated so that Tschebychev's Theorem does not apply.* This means that we don't know if the time series (7.21) has a limit, much less which limit. Hence, when histograms are constructed by sliding a window, there is no reason to expect that one has obtained either $f(z,0,T)$, where generally $f(z,0,T) \neq f_1(z,t)$. By assuming instead that one reads $z = x(0,T) = x(T)$ in the procedure, one could ask if $f_1(x,T)$ is the limit, but again, Tschebychev's Theorem fails to apply. The ensemble average is given by

$$f(x, t, t + T; z', t', t' + T) = \langle \delta(z - x(t, T))\delta(z' - x(t', T))\rangle \quad (7.23)$$

or

$$f(x, t, t + T; z', t', t' + T)$$
$$= \int \prod_{k=1}^{4} dx_k \delta(z - x_4 + x_3)\delta(z' - x_2 + x_1) f_4(x_4, t + T; \ldots, x_1, t') \quad (7.24)$$

which reduces to

$$f(x, t, t + T; z', t', t' + T)$$
$$= \int dx_3 dx_1 f_2(x_2 + z, t + T, x_3, t; x_1 + z', t' + T, x_1, t') \quad (7.25)$$

and this doesn't vanish at a rate rapid enough to yield convergence. Clearly, we cannot use Tschebychev's Theorem to argue that a long time series allows the sliding window to converge to a definite limit.

So, stationary increments or not, we cannot escape the need to construct statistical ensembles in data analysis. In particular, without the construction of an ensemble, the question of stationary vs nonstationary increments cannot even be posed, much less answered. The good news is that, with stationary increments, one could break up the time series into N "runs" where N is arbitrary. This is called "bootstrapping" in econometrics, but econometricians merely assume rather than establish increment stationarity (we analyze the common econometric assumptions in detail in Chapter 10).

A sliding window time average for the mean square fluctuation was used by Mandelbrot (1968) to analyze cotton price returns. The plot of the mean square fluctuation varied considerably (see Figure 7.2 in Mandelbrot (1968)) and did not represent scatter about a flat line. The assumption was made that the series is stationary (the series is most likely nonstationary with nonstationary increments) and the lack of convergence was assumed to mean that the increment variance is badly behaved. The variance was set equal to infinity, and a Levy distribution was deduced. Levy distributions have the fattest tails because the variance is infinite. We would expect lack of convergence of the time average of the mean square fluctuation based on the considerations above. The original conclusion that cotton returns have fat tails is therefore questionable.

7.3.3 Nonstationary increment processes

Suppose that we're given a single time series like a six-year Euro/Dollar exchange rate in returns. We can only proceed by making an ansatz of statistical repetition in the time series. To construct an approximate ensemble, we must first assume that there's a time scale on which traders' behavior is statistically repetitive. Once applied, the ansatz must be checked for correctness. We show in Section 7.4 how to apply this assumption to an FX series, and how to check it. With the time scale for repetition assumed to be one day, each day is considered as a "rerun" of the same uncontrolled trading experiment. So in a six-year time series there are about 1500 systematically repeated time series from which to contruct ensemble averages. In the next section, we show that Tschebychev's Theorem can be used to see not only that the approximate ensemble so constructed makes sense, but to tell us which quantities to measure and which to avoid. That is, in contrast with the case of repeated experiments, one cannot assume that the ensemble exists independently of the particular averages calculated. For example, we will show that it makes good sense to calculate the mean square fluctuation but not the process variance.

7.3.4 Approximate statistical ensembles

Summarizing what we've learned above, the way to test for stationary or nonstationary increments in a nonstationary process like a finance or other

7.3 Time series analysis

macroeconomic time series is via ensemble averages. Moreover, to analyze the series at all, a statistical ensemble must be constructed, otherwise no reliable analysis is possible. Sliding window time averages fail the test for Tschebychev's Theorem, so that we have no reason to believe that those averages yield correct estimates either of densities or statistical averages.

We assume here a single, historic time series. Our experience with FX time series is used as the example. Reruns of the same experiment must be replaced by evidence of statistical repetition on some time scale T_{rep}. Evidence from the FX market is shown as Figure 7.2, where $T_{rep} = 1$ day, whereas $T = 10$ minutes is the time lag for increment autocorrelations to decay to zero. This requires some explanation.

In the construction of an ensemble based on taking each day in the market as a rerun of the "trading experiment," the starting prices/returns from one day to the next are correlated. Tschebychev's Theorem can only be applied to quantities where the day-to-day correlation falls off fast enough to be negligible. FX markets run 24 hours a day, five days a week, so in our analysis the clock is arbitrarily reset to 9 am each day to define the new "run." The first return of day n at 9 am is the same as the last return of day $n-1$, and those two returns are Martingale correlated: $\langle x(t)x(t+T)\rangle = \langle x^2(t)\rangle$. Clearly, process returns are not a candidate for the application of Tschebychev's Theorem.

What about densities? The relevant correlation for the ensemble average of the one-point density is in this case $f_2(x, t+nT; y, t)$ with $T = 10$ minutes and $n = 1$ day/10 minutes $= 144$. With $f_2(x, t+nT; y, t) = p_2(x, t+nT|y, t)f_1|(y, t)$ the question is whether $p_2(x, t+nT|y, t|) \ll 1, n \gg 1$. For a diffusive process we expect this to hold independently of $x \neq y$, but getting a result close enough to zero for good convergence of the density series is highly unlikely. Correspondingly, in practice, we know that histograms based on only about 1500 points (1500 trading days in a six-year time series) have too much scatter to identify a density. In contrast with claims made on the basis of sliding-window time averages, we cannot obtain a plot with little enough scatter to identify a density even on the time scale of one day (one week or one month is far worse).

But even if densities cannot be determined empirically, various averages can be extracted pretty accurately. Consider the mean square fluctuation, the "increment variance"

$$\langle x^2(t, T)\rangle = \sum_{k=1}^{N} x_k^2(t, T) \tag{7.26}$$

at time t. There are N points in the ensemble, where $N = 1500$ for a six-year FX time series. The "independence" condition for validity of an

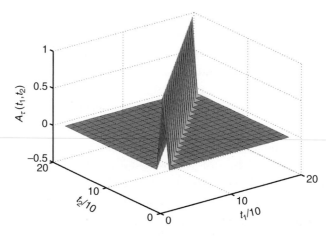

Figure 7.1 Normalized autocorrelations in increments $A_T(t_1,t_2) = \langle x(t_1,T)x(t_2,T)\rangle/(\langle x^2(t_1)\rangle \langle x^2(t_2)\rangle)^{1/2}$ for two nonoverlapping time intervals $[t_1, t_1 + T]$, $[t_2, t_2 + T]$ decay rapidly toward zero for a time lag of $T \geq 10$ minutes of trading.

ensemble average based on taking each day as a "rerun" is that $\langle x^2(t,T)x^2(t+nT,T)\rangle \approx 0$ where $nT =$ one day for FX markets. It's easy to show that

$$\langle x^2(t_1,T)x^2(t_2,T)\rangle = \langle x^2(t_2+T)x^2(t_1+T)\rangle - \langle x^2(t_2)x^2(t_1+T)\rangle \\ + \langle x^2(t_2)x^2(t_1)\rangle - \langle x^2(t_2+T)x^2(t_1)\rangle. \qquad (7.27)$$

For $t_2 \gg T$ the right-hand side vanishes pairwise linearly in T/t, so that mean square fluctuation estimates from the approximate ensemble should be pretty good. One cannot reason similarly that the process variance $\sigma^2(t) = \langle x^2(t)\rangle$ can be extracted empirically, because the required condition that $\langle x^2(t)x^2(t+nT)\rangle \ll 1$ is not met.

Summarizing, there are three time scales in the construction of the ensemble from a single long time series. First, there is the time lag T for increment autocorrelations to die out (establishing a Martingale for detrended data). For FX data this is shown as Figure 7.1, where we found that $T = 10$ minutes. Second, there is the time lag T_{corr} for day-to-day correlations to die, so that ensemble averages converge. Third, there is the time scale for behavioral repetition of the traders' T_{rep} (one day for FX trading), shown as Figure 7.2, which is the basis for the ensemble in the first place, and we clearly need $T_{corr} \leq T_{rep}$. The periodicity on which finance market ensembles are based was first noted by Gallucio et al. (1997).

Apparently, if the increments are stationary then the time scale chosen for breaking up the time series into an ensemble is arbitrary. For example, one could choose one day or any other time scale for defining reruns of the trading experiment. The limitation imposed on the choice of long time scales will be too much scatter due to too few points at each time t in the ensemble. For example, stationary increments or not, a six-year time series would require taking a time scale no longer than a day to define the ensemble. Stationary increments implies a variance linear in the time, so scaling of the one-point density with $H = 1/2$ should be checked. The advantage gained by scaling, if the data collapse can be verified, is that the intraday density can be found, and the extrapolation of that density to larger time scales of a week or a month would be a reasonable guess.

If we return briefly to the sliding window method applied to the mean square fluctuation for stationary increments, then the time average

$$\langle x^2(t,T) \rangle_{\text{timeavg}} = \frac{1}{N} \sum_{k=1}^{N} x^2(t_k, T) \qquad (7.28)$$

would meet Tschebychev's convergence requirement if $\langle x^2(t_k, T) x^2(t_{k+1}, T) \rangle \ll 1$. From the analysis above, this can be satisfied if $t_k \gg T$. In FX analysis intraday increments are strongly nonstationary and there is insufficient data to check the increments via an ensemble average calculation for interday trading. However, a "visual" inspection of Figure 7.2 below provides soft evidence that the mean square fluctuation may be linear in T for time lags $T = 1$ day. This would require $t_k \gg 1$ day in (7.27), for example, $t_k \approx 100$ days, reducing the number of data points in the sum (7.28) considerably. In the end, there is nothing to be gained from sliding the window. We see no convincing test for weakly stationary increments other than from an ensemble calculation.

Finally, ensemble averages suggest a method for detrending a time series. A trivial drift, one depending on t alone, can be removed from an increment $x(t,T)$ for each fixed (t,T). The problem of detrending a general (x,t)-dependent drift is discussed in McCauley et al. (2007c).

Lillo and Mantegna (2000, 2001) tried to define an ensemble by taking entirely different stocks as "runs of the same experiment." They used the variable $Y_k(t) = (p_k(t + T) - p_k(t))/p_k(t)$ instead of the logarithmic return, but Tschebychev's Theorem for the construction of an ensemble is not met because (i) the stocks were not detrended, and (ii) different stocks are generally pairwise correlated (Laloux et al., 1999; Plerou et al., 1999). This means, for example, there is no reason to expect that a sensible dynamic model can be extracted from the S&P 500, or from any other existing stock index; the conditions for convergence to an ensemble average simply are not met. But Lillo and Mantegna's idea of using different stocks to define an ensemble suggests an interesting alternative. The stock prices first should be detrended multiplicatively. Tschebychev's Theorem requires only bounded variances, not equal ones. One could therefore choose as an "index" a collection of detrended and pairwise uncorrelated stocks. Such an index can be expected to converge to an ensemble averages if the number of stocks is large enough. Clearly, 500 detrended and pairwise uncorrelated stocks would be far too few; we will show below that we would need at least 1500, and preferably many more than that.

7.4 Deducing dynamics from time series

We now describe our study of a six-year time series of Euro/Dollar exchange rates from Olsen and Associates recorded at one-minute intervals. We started by assuming that one day is the time scale for repetitive behavior. That provided the ensemble of 1500 days from which the averages shown in Figure 7.1 and 7.2a were calculated. We verified the assumption of statistical

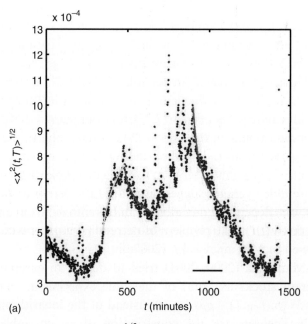

(a)

Figure 7.2a The rmsf $\langle x^2(t,T)\rangle^{1/2}$ of the daily Euro-Dollar exchange rate is plotted against time of day t, with time lag $T = 10$ minutes to ensure that autocorrelations in increments have died out (Figure 7.1). This shows that the increments are nonstationary and depend strongly on starting time t. Both of the plots 7.2a and 7.2b would be flat were the increments $x(t,T)$ stationary. The plot represents approximately the (square root of the) average diffusion coefficient. The four lines drawn in the plot represent regions where scaling with four different Hurst exponents can be used to fit the data.

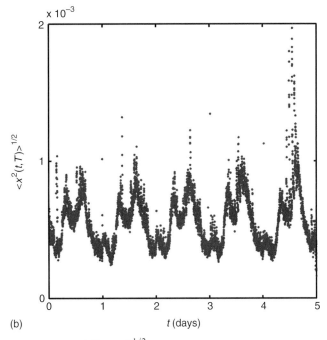

Figure 7.2b The rmsf $\langle x^2(t,T)\rangle^{1/2}$ plotted for five successive trading days. Note that the same intraday average is repeated during each trading day, providing the statistical regularity necessary as the basis for analyzing a nonstationary time series via the construction of an approximate ensemble.

repetition of behavior as Figures 7.2a and 7.2b, showing that the daily rmsf in returns is repeated for each trading day of the week, with the scatter somewhat worse on Fridays, perhaps because this is the last trading day. The valleys in Figure 7.2 may reflect lunch breaks and other daily regularities in the life of a trader; the peaks represent times of greatest activity (one could try to correlate this with volatility; see also Cross (1998)). It would have been impossible to extract the knowledge we've obtained had we relied on standard methods of econometrics and statistical analysis, on regression methods, for example. The discussion in this chapter can be understood as the suggested replacement for econometrics and standard statistical methods in macroeconomics and beyond.

Summarizing, we performed ensemble averages for each time t within a day based on the assumption that each day is a rerun of the same process. That means, for a six-year time series, the averages for each time of day t were calculated on the basis of about 1500 points, there being about 1500 days in six years. Using log return levels $x(t) = \ln(p(t)/p_c)$, a very small drift was removed initially from the data at each time t. And because a trading day runs for 24 hours, we reset the clock each morning at 9 am. The data represent a single FX market; we did not mix statistics from different markets (say, New York and London) together.

That the intraday level differences $x(t,T) = x(t + T) - x(t) = \ln(p(t + T)/p(t))$ are strongly nonstationary is shown by the mean square fluctuation $\langle x^2(t,T) \rangle$, which varies considerably as the time t is increased for fixed lag time $T = 10$ minutes (Figure 7.2).

We've shown in Chapter 3 that the lack of increment autocorrelations guarantees Martingale differences

$$x(t,T) = \int_t^{t+T} b(x(s), s) dB(s) \tag{7.29}$$

so that a diffusion coefficient $D(x,t) = b^2(x,t)$ describes the traders' behavior. Can we discover the diffusion coefficient that characterizes the FX market Martingale? This is a problem not less demanding than discovering the one-point density empirically. We were not able to discover either quantity, but we can say something about both. In particular, Figure 7.2 can be understood as the ensemble average of the diffusion coefficient.

The reason that Figure 7.2a represents an *unconditioned* average follows from the lack of control of any starting point $x(t)$ at "opening time" each day. Figure 7.2a was constructed as follows: for each time interval $[t, t + T]$ neither end point $x(t)$ or $x(t + T)$ is controllable from one day to the next over the 1500 days in the sample. That is, unlike in a laboratory, we could not rerun the experiment by choosing the same initial condition at each starting time. The best we could do was to calculate the ensemble average of the difference $x(t,T)$ for fixed t and T using the 1500 days. With T fixed we redo the calculation for each time t during the day, generating Figure 7.2a. Here's the corresponding theoretical average. Setting $z = x(t,T)$, the density

$$f_{\text{incr}}(z, t, T) = \int dx p_2(x + z, t + T | x, t) f_1(x, t) \tag{7.30}$$

describes the increments and can be used to calculate

$$\langle z^2 \rangle = \iint dx dz z^2 p_2(x + z, t + T | x, t) f_1(x, t) \tag{7.31}$$

For $t \gg T$, using the definition of the diffusion coefficient

$$D(x,t) \approx \frac{1}{T} \int_{-\infty}^{\infty} dy (y - x)^2 p_2(y, t + T | x, t), \quad T \ll t \tag{7.32}$$

yields the quantity measured by us in Figure 7.2 as would be predicted by

$$\langle x^2(t,T) \rangle \approx T \int \mathrm{d}x D(x,t) f_1(x,t) \tag{7.33}$$

where $D(x,t)$ characterizes the traders' behavior during a single trading day and f_1 is the corresponding one-point density. Figure 7.2 is therefore not a volatility, which would require a conditioned average, it's simply the unconditioned ensemble average of the diffusion coefficient. *There is noise in Figure 7.2 because there are only 1500 points for each time t, otherwise the plot should be piecewise-smooth*, allowing that the diffusion coefficient may have piecewise discontinuous slope. We can deduce a diffusion coefficient and one-point density only for the time intervals in Figure 7.2a where a data collapse due to scaling can be used roughly to fit the mean square fluctuation. See van Kampen (1981) for an early discussion of conditions for discovering the diffusion coefficient empirically for stationary processes.

The four lines drawn into Figure 7.2a represent time intervals where we could fit the data via a scaling function with a Hurst exponent H. The Hurst exponent is different for each region, but $H \approx 0.35$ was necessary to get a data collapse for the one-point density for the longest line, the line based on the most data points. We now show how to deduce the corresponding diffusion coefficient for that time interval.

Within the largest interval shown in Figure 7.2a, a data collapse $F(u) = t^H f(x,t)$, $u = x/t^H$ with $H \approx 0.35$ can be used to fit the density for the longest line at different times of day t. Figure 7.3a shows that the scaling function $F(u)$ has no fat tails, is instead approximately a two-sided exponential. We have Martingale dynamics, and the transition density obeys

$$\frac{\partial p_2(x,t|x_0,t_0)}{\partial t} = \frac{1}{2} \frac{\partial^2}{\partial x^2} (D(x,t) p_2(x,t|x_0,t_0)) \tag{7.34}$$

with $p_2(x,t|x_0,t) = \delta(x - x_0)$, and "local volatility" is described by the inseparable (x,t)-dependence in the diffusion coefficient $D(x,t)$. The one-point density obeys the same pde

$$\frac{\partial f_1(x,t)}{\partial t} = \frac{1}{2} \frac{\partial^2}{\partial x^2} (D(x,t) f_1(x,t)) \tag{7.35}$$

Figure 7.3a shows the data collapse that we fit using

$$f_1(x,t) = t^{-H} F(u), \quad u = x/|t|^H \tag{7.36}$$

We then obtain from (7.35) that

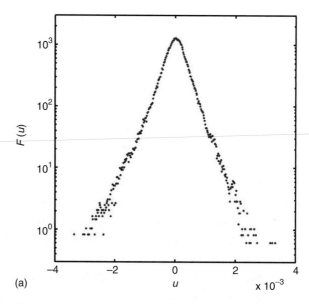

Figure 7.3a This data collapse for H = 0.35 was obtained for the longest line shown in Figure 7.2a, where 10 min. $\leq T \leq$ 160 min. Note that $F(u)$ is slightly asymmetric and is approximately exponential, showing that the variance is finite and thereby ruling out Levy densities.

$$2H(uF)' + (\widehat{DF})'' = 0 \tag{7.37}$$

where

$$D(x,t) = |t|^{2H-1}\widehat{D}(u) \tag{7.38}$$

From this, we obtain

$$\widehat{D}(u)F(u) = \widehat{D}(0)F(0) - 2H \int_0^u uF(u)du \tag{7.39}$$

For a two-sided exponential density

$$F(u) = \begin{cases} A_+ e^{-\nu u}, u > 0 \\ A_- e^{\gamma u}, u < 0 \end{cases} \tag{7.40}$$

we obtain

$$\widehat{D}(u) = \begin{cases} 2H[1 + \nu u]/\nu^2, u > 0 \\ 2H[1 - \gamma u]/\gamma^2, u < 0 \end{cases} \tag{7.41}$$

It is this diffusion coefficient that generates the Martingale dynamics of a scaling exponential density.

We could calculate the transition density p_2 numerically, but not analytically. We understand the dynamics of the nonstationary variable diffusion processes that describe FX markets, trading produces a Martingale in

7.6 Volatility measures

detrended returns, and Martingales are diffusive. Volatility is caused by the variable diffusion coefficient $D(x,t)$, which is in turn caused by the traders' nonstationary intraday behavior. Volatility and instability are not necessarily the same idea.

FX market instability is characterized as follows. First, we have a Martingale process (uncorrelated increments). The mean square fluctuation does not stabilize on the time scale of a day so that the variance does not approach a constant. Considering unconditioned Martingale averages

$$\sigma^2(t+T) = \langle x^2(t,T) \rangle + \sigma^2(t) \tag{7.42}$$

if we take the time lag T to be one day instead of 10 minutes, then we see from Figure 7.2b that the mean square fluctuation *visually appears* to be approximately t-independent, so that the increments may be approximately weakly stationary for time lags of $T \approx 1$ day. That is, the variance may be roughly linear in the time on the time scale of a day or longer, which is expressed in (7.42) as $\sigma^2(t) \approx t \langle x^2(1) \rangle$, $\langle x^2(t,T) \rangle \approx T \langle x^2(1) \rangle$ for $t \approx T \approx 1$–5 days. In other words, for time scales on the order of a day or more, the variance may increase linearly with the time. *There is no evidence that the variance approaches a constant for any measurable time scale.* Another way to say it is that finance markets are unstable.

7.5 Early evidence for variable diffusion models

The earliest indication of the need for variable diffusion in returns was the idea of "implied volatility" in the Black–Scholes model, although implied volatility was never understood in that light until our 2003 paper (McCauley and Gunaratne, 2003).

In the Black–Scholes model, the constant σ_1 in the Gaussian transition density is required to vary with strike price, which is not allowed by the model. In that case the observed option price is used and the "volatility" is then implied. This led to the notion of variance as volatility; the variance $\sigma^2(t) = ct$ predicted by the Gaussian returns model is wrong. In that model the increments are stationary. A more accurate statement is that the prediction for the mean square fluctuation $\langle x^2(t,T) \rangle = \langle x^2(T) \rangle = cT$ is wrong. We've seen above how this fails. First, the increments are not stationary and, second, the one-point density is not Gaussian.

7.6 Volatility measures

Various volatility measures have been proposed in the literature. We can take as one volatility measure the conditioned mean square fluctuation, chosen in the 1980s by Engle (who did not study Martingales). For Martingales the conditioned average yields

$$\langle x^2(t,T)\rangle_{\text{cond}} = \int_t^{t+T} ds \int_{-\infty}^{\infty} dy D(y,s) p_2(y,s|x,t) \qquad (7.43)$$

which depends on the last observed point x at time t. The unconditioned average is

$$\langle x^2(t,T)\rangle = \int_t^{t+T} ds \int_{-\infty}^{\infty} dy D(y,s) f_1(y,s) \qquad (7.44)$$

and is shown as Figure 7.2. In the nonsystematic modeling of finance data ("stochastic volatility" is an example) "volatility" is often modeled separately from returns. This is inconsistent. Given a stock, there is only a single, historic nonstationary price series from which the time series for log returns is directly obtained. From that series all calculations of volatility as one choice of correlation or another must follow, by self-consistency. Otherwise stated, volatility is simply one way of discussing fluctuations in returns and there is only one time series for returns. Stochastic volatility models are seen as unnecessary and inconsistent from this standpoint.

In the literature on autoregressive conditional heteroskedasticity (ARCH) and generalized ARCH (GARCH) processes (Chapter 10), it's normally assumed that volatility goes hand in hand with local nonstationarity, but that in the long run finance time series are stationary. There's no evidence at all for long time stationarity, and we'll show in addition that ARCH and GARCH models generalized to nonstationary processes violate both the empirical data and the EMH: the increment autocorrelations in those models cannot vanish.

7.7 Spurious stylized facts

The purpose of this section is to make clear the misconceptions that can arise from using a sliding window to build histograms from a time series with nonstationary increments.

We can begin with Hommes's (2002) statement of the "observed stylized facts" of FX markets: (i) "asset prices are persistent and have, or are close to having, a unit root and are thus (close to) nonstationary"; (ii) "asset returns are fairly unpredictable, and typically have little or no autocorrelation"; (iii) "asset returns have fat tails and exhibit volatility clustering and long memory. Autocorrelations of squared returns and absolute returns are

7.7 Spurious stylized facts

significantly positive, even at high-order lags, and decay slowly." These three statements, although largely wrong, reflect a fairly standard set of expectations in financial economics. Next, we contrast those expected stylized facts with the hard results of our FX data analysis.

In point (i) above, "unit root" means that in $p(t+T) = \lambda p(t) + noise$, $\lambda = 1$. That's simply the necessary condition for a Martingale, and rules out persistence like fBm but also stationarity. In the focus on a unit root, economists are searching for evidence of a stationary time series, requiring $0 < \lambda < 1$ (see Chapter 10). Prices are not "close to nonstationary," prices are very far from both strong and weak stationarity. (ii) Increment autocorrelations in FX market returns will vanish after about 10 minutes of trading. By "persistence" Hommes presumably means serial correlations, but he should have noted that a continuous coordinate transformation $x(t) = \ln(p(t))$ cannot possibly erase pair correlations. Both detrended prices and detrended returns have (Martingale) positive serial correlations, e.g. with $x(0) = 0$. The autocorrelations in increments approximately vanish after 10 minutes of trading (Figure 7.1), $\langle x(t+T)x(t) \rangle = \langle x^2(t) \rangle > 0$. (iii) We find no evidence for fat tails in intraday trading (Figure 7.4a), and no evidence for Hurst exponent scaling persisting on the time scale of a day (Figure 7.2a). We offer no comment on the question of necessity of memory to understand volatility clustering at this point; we note only that the claim has not been proven.

Our main point in this section is: *the data analyses used to arrive at the expected stylized facts have generally used a technique called "sliding windows."* The aim of this section is to explain that sliding windows can produce spurious, misleading results because a sliding window presumes stationarity of the increments. Stated in the language of econometrics, the differences between levels are not stationary; the intraday differences are strongly nonstationary (Figure 7.2a). Only one previous FX data analysis that we are aware of (Gallucio et al., 1997) showed that sliding windows lead to predicting a Hurst exponent $H_s = 1/2$, even if the original time series would scale with $H \neq 1/2$. That analysis correctly identified the cause as nonstationarity of the increments. We will explain theoretically why sliding a window on nonstationary increments yields $H_s = 1/2$.

It must first be realized that there are three separate one-point densities. First, there is the empirically correctly obtained density $f_1(x,t)$. Second, there is the increment density $f_{incr}(z,t,T) \neq f_1(x,t)$ (7.17) because the increments are nonstationary. Third, there is the spurious density $f_s(z,T)$ obtained from a sliding window analysis (from an empirical time average over the time t with lag time T fixed), which equals neither of the first two densities and cannot be calculated analytically because the ergodic theorem cannot be applied.

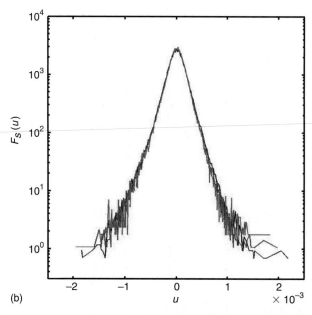

Figure 7.3b The "sliding interval scaling function" $F_s(u_s)$, $u_s = x_s(T)/T^{H_s}$, is constructed using a sliding window time average for the same interval as in Figure 7.3a for $T = 10$, 20, and 40 min. Note that fat tails have been generated spuriously by the sliding window, and that a Hurst exponent $H_s = 1/2$ has been generated contradicting the fact that the correct scaling function shown as Figure 7.3a has $H = 0.35$.

Nearly all previous analyses (Osborne, 1964; Mantegna and Stanley, 1995, 1996; Friedrich *et al.*, 2000; Dacorogna *et al.*, 2001; Borland, 2002; di Matteo *et al.*, 2003; McCauley and Gunaratne, 2003) have used a time average called a sliding window. To illustrate the spurious stylized facts generated by constructing time averages using a sliding window, we apply that method to a time series with uncorrelated nonstationary increments, with no fat tails and with a Hurst exponent $H \neq 1/2$, namely, a time series generated by the exponential density (7.40) with $H = 0.35$ and linear diffusion (7.41). The process is Markovian. Figure 7.4a was generated by taking 5,000,000 independent runs of the Ito process, each starting from $x(0) = 0$ for $T = 10$, 100, and 1000. The sliding window result is shown as Figure 7.4b. In this case, the sliding windows appear to yield a scale-free density $F_s(u_s)$, $u_s = x_s(T)/T^{H_s}$, from an empirical average over t that one cannot even formulate analytically, *because for a nonstationary process there is no ergodic theorem*. Not only are fat tails generated artificially here, but we get a Hurst exponent $H_s = 1/2$ that disagrees with the Hurst exponent used to generate the time series. This is the method that has been used to generate stylized facts in nearly all existing finance data analyses. Figure 7.3b shows how the approximately exponential

7.7 Spurious stylized facts

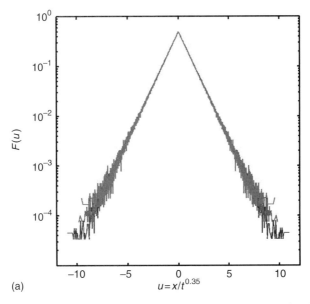

(a)

Figure 7.4a The scaling function $F(u)$ is calculated from a simulated time series generated via the exponential model, $\widehat{D}(u) = 1 + |u|$ with $H = 0.35$. An ensemble consisting of 5,000,000 independent runs of the exponential stochastic process was used here.

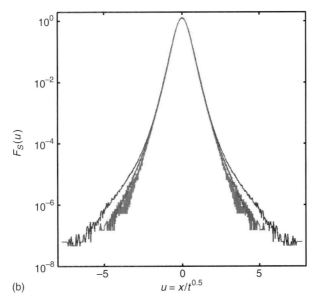

(b)

Figure 7.4b The "sliding window scaling function" $F_s(u_s)$, $u_s = x_s(T)/T^{H_s}$ was calculated by using a time average on the same simulated data. Note that F_s has fat tails whereas F does not, and that $H_s = 1/2$ emerges, contradicting the fact that $H = 0.35$ was used to generate the original time series. That is, the time average generates two spurious stylized facts.

FX density of Figure 7.3a is fattened by sliding a window on the six-year Olsen and Associates time series. We can explain the origin of $H_s = 1/2$.

I offer here an anecdote that shows that people are not necessarily consistent in their viewpoints (Holton, 1993). Gunaratne used the sliding window method to deduce the exponential density for FX returns that was presented in the first edition of this book. In 2005, in an attempt to understand what Hurst exponents had to do with correlations, and to try to understand claims made in the literature about "nonlinear Fokker–Planck equations" (Borland, 2002; Frank, 2004; McCauley, 2007), our group (Bassler, Gunaratne, and McCauley) first noticed what Mandelbrot and van Ness (1968) had made clear but that nearly everyone had since forgotten: the argument that $H \neq 1/2$ implies long time correlations is based on an implicit assumption of stationary increments. We then discovered a class of Markov models that scale (Bassler et al., 2005) and provided examples. We also showed how the predictions of "nonlinear Fokker–Planck pdes" are only superficially nonlinear, and are actually generated by a linear diffusion model. We explained that, when increment autocorrelations vanish, $H \neq 1/2$ is simply a signal that increments are nonstationary. We then proceeded to analyze our FX data using a sliding window! Bassler found that we could not really fit the fat tails in the sliding window density by any number of terms combined as polynomials in x/t^H in the diffusion coefficient. Eventually, Gunaratne noted: we've explained that the increments should be nonstationary, so we can't slide a window. That illustrates just how inconsistent a viewpoint can be, even in mathematics. One understands something, and one assumes something else that contradicts the original assertion. As Holton notes, this "inconsistency of worldview" is probably quite common in humans. We return next to our technical discussion.

With uncorrelated, nonstationary increments, in an interval where scaling fits, the unconditioned mean square fluctuation is

$$\langle x^2(t,T)\rangle = \left\langle (x(t+T)-x(t))^2 \right\rangle = \langle x^2(1)\rangle[(t+T)^{2H} - t^{2H})] \qquad (7.45)$$

In most existing data analyses we generally have $T/t << 1$ (where $T = 10$ minutes and t ranges from opening to closing time over a day), so

$$\langle x^2(t,T)\rangle \approx \langle x^2(1)\rangle 2Ht^{2H-1}T \qquad (7.46)$$

A sliding window then (illegally, because the process is not ergodic so that time averages cannot be replaced by ensemble averages) averages empirically over t,

$$\langle x^2(t,T)\rangle_s \approx \langle x^2(1)\rangle 2H\langle t^{2H-1}\rangle_s T \qquad (7.47)$$

yielding $\langle x^2(t,T) \rangle \approx T^{2H_s}$ with $2H_s = 1$. Sliding window Hurst exponents $H_s = 1/2$ have been reported in the literature (Fogedby et al., 1992; Borland, 1998), but without any correct explanation of how they arise from selfsimilar models where increments are uncorrelated with $H \neq 1/2$. That $H_s = 1/2$ is a consequence of using sliding windows was first reported by Gallucio et al. in 1997.

7.8 An sde for increments?

We've argued that $x(t)$ is always a well-defined variable in an sde but that the increment $x(t,T)$ is not. This must be qualified.

If we consider t as the variable, with T fixed, then we cannot use Ito's lemma to derive an sde for $x(t,T)$. At best we would obtain three coupled sdes for $x(t,T)$, $p(t)$, and $p(t+T)$.

If we fix t and let T vary, then we can derive an sde. This makes sense because t is an initial time; T is the real time variable in the stochastic process. Denote $t = s + T$ with s fixed. Then $p(s+T)$ is the variable in $x(t,T)$, $p(s)$ is a fixed reference price, and from the sde for $p(s+T)$ with T varying we obtain

$$dx(t,T) = (\mu - \check{D}(x(s, s+T), s+T))dT + \sqrt{\check{D}}dB(T) \qquad (7.48)$$

since B is time-translationally invariant. If one looks at a scaling model then we see easily that the diffusion coefficient does not scale in T, it scales in $s + T$, and setting $T = 0$ produces neither the price at which $\check{D}$ has a minimum, nor the consensus price. That is, this sde and the corresponding Fokker–Planck pde are cumbersome to work with. In general, the diffusion coefficient $\check{D}$ also depends on the consensus price p_c, if not on two hidden price scales.

So we do not know at this stage how to apply the increment sde in any empirically useful way. This concludes our presentation of empirically based modeling via ensemble averages.

7.9 Topological inequivalence of stationary and nonstationary processes

In rudimentary applications of regression analysis it's sometimes assumed that a nonstationary time series can be transformed into a stationary one. Such a transformation is trivially true locally but is generally impossible to construct globally. Stationarity is an analog of the notion of "integrability" in nonlinear dynamics (Bassler et al., 2008). We show next that global transformations from nonstationarity to stationarity are far from guaranteed.

Locally seen, every sde is a Wiener process (the noise is always locally white): with

$$dx = R(x,t)dt + \sqrt{D(x,t)}dB \tag{7.49}$$

the local solution, meaning the solution over a *very* short finite time interval $\delta t = t - t_0$, is

$$x(t) \approx x_0 + R(x_0, t_0)\delta t + \sqrt{D(x_0, t_0)}\delta B \tag{7.50}$$

With the transformation $y = (x - x_0)/(\sqrt{D(x_0,t_0)})\delta t$ we get a stationary process: $\langle y^2 \rangle = 1, \langle y \rangle = 0$, and the density of y is a stationary Gaussian (see also www.xycoon.com/non_stationary_time_series.htm and related papers on regression analysis, which go no further than this). Next, we ask if such a transformation is globally possible. As in nonlinear dynamics or differential geometry, this is an integrability question.

The integrability problem (first addressed by Giulio Bottazzi) can easily be formulated by using Ito calculus. Starting with the sde for $x(t)$, we ask for a global transformation $y = G(x,t)$ to a Wiener process. From a Wiener process $B(t)$, one can trivially transform to a stationary process $B(1) = t^{-1/2}B(t)$. That is, the scale-invariant part $F(B/t^H) = t^H f_1(B,t)$ of the Gaussian density is stationary in the rescaled variable $u = B/t^H$. Given the transformation $y = G(x,t)$,

$$dy = \frac{\partial G}{\partial t}dt + \frac{D(x,t)}{2}\frac{\partial^2 G}{\partial x^2}dt + \frac{\partial G}{\partial x}\sqrt{D(x,t)}dB \tag{7.51}$$

the condition for a Wiener process is

$$\begin{aligned}\frac{\partial G}{\partial X}\sqrt{D(X,t)} &= \sigma(t) \\ \frac{\partial G}{\partial t} + \frac{D(X,t)}{2}\frac{\partial^2 G}{\partial X^2} &= \mu(t)\end{aligned} \tag{7.52}$$

The required integrability conditions (the conditions under which G exists globally) are

$$\frac{\partial^2 G}{\partial x \partial t} = \frac{\partial^2 G}{\partial t \partial x} \tag{7.53}$$

with

$$\begin{aligned}\frac{\partial G}{\partial t} &= \mu(t) - cR/\sqrt{D} + \frac{1}{4}\frac{\partial D}{\partial x}/D^{3/2} \\ \frac{\partial G}{\partial x} &= c/\sqrt{D}\end{aligned} \tag{7.54}$$

An easy calculation shows that, aside from the lognormal process, the only process satisfying global integrability is another Wiener process, $y = \mu t + cB$

(McCauley et al., 2007c). A nonstationary process with $D(x,t)$ depending on x generally cannot be transformed to a Wiener process. Processes with R and D depending *only* on t are trivially Wiener by a simple transformation of variables.

One can ask more generally if a nonstationary process can be transformed into an asymptotically stationary process like OU. This question can also be formulated as an integrability question, and there is at this stage no general answer. Given some asymptotically stationary process

$$dy = -\gamma(y)dt + \sqrt{E(y)}dB \tag{7.55}$$

with the appropriate conditions for stationarity on γ and E, the conditions are then

$$\frac{\partial G}{\partial t} + \sigma_1 R/\sqrt{D} - \frac{1}{4}\frac{\partial D}{\partial x}/D^{3/2} = -\gamma(y)$$
$$\frac{\partial G}{\partial x}\sqrt{D} = E(y) \tag{7.56}$$

where we must know G in advance and then invert to obtain $x = H(y,t)$ in order to test for integrability. No general theory is available, and our conjecture is that the procedure is generally impossible. The deterministic analog would be that nonintegrable deterministic systems cannot be transformed into integrable ones. In any case, there is no reason to believe *a priori* that an arbitrary nonstationary process can be transformed into a stationary one.

A scaling one-point density can be transformed into a stationary one-point density, $F(u) = t^H f_1(x,t)$. However, both the transition density p_2 (which generally does *not* scale) and the Ito sde show that the stochastic process studied in the variable u is nonstationary. So an arbitrary scaling process cannot be transformed into a stationary one.

This eliminates the assumption that nonstationary time series can be transformed into stationary ones. But mathematical economists are far more sophisticated than this naive assumption (7.50). On the subject of "Integration I(d)" and cointegration, to be covered in Chapter 10, the claim is made that nonstationary levels can be made stationary by taking differences.

8
Martingale option pricing

8.1 Introduction

Betting is risky, and for noise traders financial markets are formalized gambling casinos. The idea of a bet is to take a risk in order to try for a big win. A hedge on the bet reduces the risk, reducing both the possible win and the possible loss. Buying stocks in both rain and beach umbrellas reflects the idea of a hedged bet. Options provide a more direct way of hedging a bet on a stock, bond, or FX. Even money, however, is risky, as inflation can occur and a currency can be degraded systematically by the policies of the government in charge.

A stock, bond, or a foreign currency is a risky paper "asset" because the price fluctuates freely against your local currency. A bank deposit in the local currency, CD, or money market account is called "risk free" in financial math texts. Obviously, that idealization ignores the riskiness of the local currency (which reflects a nation's financial and fiscal policies) against necessary imports like oil and food. The riskiness of the Dollar as the world's default reserve currency is discussed in Chapter 9. Here, we will write as if a local currency could be "risk-free." We will ignore inflation and consider only local bank interest rates. In truth, because of market instability, nothing in finance is risk-free.

A bond is a loan at a fixed interest rate, and fluctuates in price in anticipation of changes in future money market interest rates. A stock guarantees the owner nothing definite in returns, whereas a bond is guaratreed to pay back the principle plus interest if held to maturity, if the issuer doesn't go belly-up beforehand. Following Black and Scholes (1973), a stock can be understood as an option on real assets at an indefinite future date. So a stock option is an option on an option. A local currency can change in value compared with a foreign currency. This is of much practical interest because a local currency generally must be converted into a foreign currency or equivalent to pay for imports, or to be paid for exports. Hedges of one currency against another are used in the attempt to limit risk in contracts for future delivery or purchase of goods. To keep the language simple we will often call stocks, bonds, and FX "stocks" in what follows.

A call is a contract that gives one the right, but not the obligation, to buy a stock at a predetermined price K within a time frame $[t,T]$ where t is present time and T is the expiration date of the contract. The price K is called the strike price. Owning a put gives one the right to sell a stock at a predetermined price K in the same time interval. With a good enough credit rating and evidence of enough money to play, one can obtain from a broker the right to trade puts and calls. Merely owning enough shares of a stock generally confers the right to sell covered calls on those shares (a covered call is a call on the number of shares of a stock that you own). More precisely, we have described a so-called American option. A so-called European option (Chapter 5) can be exercised only at the expiration time T. This type of option is of less practical interest but is quite easy to formulate mathematically as an initial value problem, and so is of pedagogic interest.

The aim here is to formulate mathematically the expression for a fair price of a European option. Given the fair price, if an option sells above or below that price then one can say that the option is overvalued or undervalued, and that would allow one to define arbitrage opportunities. There is one catch: the predicted fair price will implicitly assume a normal liquid market. We don't know how to price options meaningfully in a crash because a crash is a surprise, and the liquid market fair price is formulated explicitly on the assumption that the future will be the same statistically as the past (the reader is advised to review Section 4.13 at this point).

At expiration time T the fair price is easy to formulate. Consider a call option. Denote the expected fair price by C. If the asset price at expiration satisfies $p(T) < K$ then the call is worthless, $C = 0$, because I can buy the stock at time T more cheaply than is specified in the contract. Suppose that $p(T) > K$ by an amount that is greater than transaction costs and taxes. Then one should exercise the option because there is an immediate arbitrage opportunity: one can buy the stock at price K and immediately sell it at price $p(T)$ *if the market is liquid*. Ignoring brokerage fees and taxes, the fair price is then given by

$$C = \max[p(T) - K, 0] = (p(T) - K)\theta(p(T) - K) \tag{8.1}$$

Normally, one never owns the stock in an options trade; a covered call provides an exception. Generally, the discount brokerage contract takes care of all buying and selling in real time for the stock market. Small fish ($10,000–100,000) are more or less barred from the pond of currency options because in that case discount brokers don't exist, and US banks act on small accounts bureaucratically at snail's pace.

The idea of "fair option pricing" is to use it as a benchmark to look for "mispricing" to trade on. To be confident, one would need first to establish that the predicted fair option price accurately describes real prices in normal liquid markets.

8.2 Fair option pricing

8.2.1 What is a fair price?

The question of "fairness" in mathematics and in life is not unique (Davis and Hersch, 2005). By a fair price in finance theory is meant that the effective interest rate on a portfolio equals the bank, or "risk neutral," interest rate. This provides the basis for arbitrage in normal, liquid markets: if the observed portfolio interest rate differs from the bank interest rate, then one might interpret this as a buy or sell signal. We will derive the option price from a portfolio that increases in value at the bank interest rate, and then prove that that option price satisfies a Martingale condition.

8.2.2 A phenomenological model

Before getting down to serious mathematics, we first entertain the reader with an example of the kind of phenomenological reasoning that's been found useful in physics (McCauley and Gunaratne, 2003). The main question for us is: what is a fair price for an option at present time $t < T$, where $p = p(t)$ is known and $p(T)$ is unknown (this obviously requires a conditional average over $P(T)$). Here's how a physicist might reason. We can extrapolate our expectation (8.1) by averaging over what we don't know, namely, $p_T = p(T)$. This requires *assuming* that the empirically determined transition density $g_p(p_T, T|p, t)$ at future time T is the same as the one that is known for times up to the present time t, because we have no scientific way to construct this density other than by using existing statistics. That is, we assume that the future will be statistically the same as the past, that there will be no surprises in the market. If the market has no memory, then we have a Markov process with price Green function $g_p(p_T, T|p, t)$ satisfying the Fokker–Planck pde for the stock price process. Given that $g_p(p, t|p_0, t_0)dp = g_p(x, t|x_0, t_0)dx$ where g is the returns Green function, and taking into account the time value of money in the bank, then we arrive at the prediction

8.2 Fair option pricing

$$C(p,K,T-t) = e^{-r_d(T-t)}\langle (p(T) - K)\vartheta(p(T) - K)\rangle$$

$$= e^{-r_d(T-t)} \int_{\ln K/p_c}^{\infty} (p(T) - K)g(x_T,T|x,t)\mathrm{d}x_T \qquad (8.2)$$

where r_d is on the order of the bank interest rate and $x = \ln(p/p_c)$ where p_c is the consensus price. This is the fair price estimate that a physicist would expect (Gunaratne, c. 1990). The problem here is to find the transition density that describes the market correctly, which is nontrivial.

For a put we correspondingly obtain

$$P(p,K,T-t) = e^{-r_d(T-t)}\langle (p(T) - K)\vartheta(p(T) - K)\rangle$$

$$= e^{-r_d(T-t)} \int_{-\infty}^{\ln K/p_c} (p(T) - K)g(x_T,T|x,t)\mathrm{d}x_T \qquad (8.3)$$

Note that we can get rid of the stock "interest rate" R, which is hard or impossible to know accurately, in the Fokker–Planck pde by imposing a "working man's Martingale condition" (Gunaratne, c. 1990)

$$\langle p(t) \rangle = p^{r_d t} \qquad (8.4)$$

thereby fixing R by the "cost of carry" r_d, which traders take to be a few percentage points above the bank interest rate r.

Next, if we make the assumption that present time prices don't fluctuate far from the consensus price p_c in a normal liquid market, $p \approx p_c$, then with $x(t) \approx 0$ we obtain the approximation

$$C(p,K,T) \approx e^{-r_d T} \int_{\ln K/p}^{\infty} (p(T) - K)f_1(x_T,T)\mathrm{d}x_T \qquad (8.5)$$

where $f_1(x,t) = p_2(x,t|0,0)$ is the empirically observed density of returns. This prediction was first written down and used by Gemunu Gunaratne in 1990 to price options successfully using the exponential density (McCauley and Gunaratne, 2003; McCauley, 2004). Next, we return to our formal development in order to derive a more rigorous notion of fair option pricing. In the end, the phenomenological prediction (8.5) can be used in practice.

8.2.3 The delta hedge strategy

It's easy to show that the delta hedge strategy, when based on a nontrivial diffusion coefficient $D(x,t)$, is still instantaneously "risk-free," just as in the case of the Black–Scholes–Merton model based on Gaussian returns, where $D =$ constant

$$rw = \frac{\partial w}{\partial t} + rp\frac{\partial w}{\partial p} + \frac{d(p,t)p^2}{2}\frac{\partial^2 w}{\partial p^2} \qquad (8.6)$$

With the transformation from prices to returns, $x = \ln p$, the option price is a scalar and obeys $u(x,t) = w(p,t)$ so we get

$$ru = \frac{\partial u}{\partial t} + (r - D(x,t)/2)\frac{\partial u}{\partial x} + \frac{D(x,t)}{2}\frac{\partial^2 u}{\partial x^2} \qquad (8.7)$$

Using the time-transformation

$$u = e^{r(t-T)}v \qquad (8.8)$$

equation (8.7) becomes

$$0 = \frac{\partial v}{\partial t} + (r - D/2)\frac{\partial v}{\partial x} + \frac{D}{2}\frac{\partial^2 v}{\partial x^2} \qquad (8.9)$$

The pde is exactly the backward-time equation, or first Kolmogorov equation, corresponding to the Fokker–Planck pde for the market Green function (transition density) of returns g if we choose $\mu = r$ in the latter. With the choice $\mu = r$ both pdes are solved by the same Green function so that no information is provided by solving the option pricing pde (8.9) that is not already contained in the Green function of the market Fokker–Planck equation. Of course, we must interpret (x,t) in (8.9) as the initial data for the Fokker–Planck pde, as $v = g(x_T, T|x, t)$.

We can now use the market Green function to price options:

$$C(p, K, T-t) = e^{r(t-T)}\int_{-\infty}^{\infty}(p_T - K)\theta(p_T - K)g(x_T, T|x, t)dx_T \qquad (8.10)$$

where $x_T = \ln(p_T/p_c)$ and $x = \ln(p/p_c)$ where p is the observed price at present time t. In the delta hedge model financial purists take the arbitrary interest rate r to be the risk-free (bank or CD) rate, but traders do not necessarily follow that advice.

There's a subtle point that should be mentioned. Although the option price transforms like a scalar, the transition density g in (8.10) transforms like a density. If we transform to price variables under the integral sign then we must use $gdx = g_p dp$ where g_p is the price transition density that solves (8.9) for the delta function initial condition. The best way to see this is that, in (8.10), we average over the initial price p with density g.

If we restrict to $x = 0$, so that $p = p_c$, then this is essentially the formula we used to price options empirically. It means that we have approximated an arbitrary stock price p by the consensus price p_c. That this doesn't get us into trouble indicates option pricing is not very sensitive to some details. Indeed, option pricing is not a strong test of the correctness of an underlying model of market dynamics. This much was covered in McCauley (2004). The reference for the next section is McCauley *et al.* (2007b). It proves for arbitrary diffusion coefficients (actually restricted to quadratic growth or less, to ensure continuity of the stochastic process) what Harrison and Kreps (1979) proved for the Black–Scholes model.

8.2.4 The martingale condition

We can show that the generalized Black–Scholes pde above is equivalent to a Martingale in the appropriately discounted stock price. The Black–Scholes pde is equivalent via a time-transformation to the backward-time Kolmogorov pde

$$0 = \frac{\partial v}{\partial t} + (r - D/2)\frac{\partial v}{\partial x} + \frac{D}{2}\frac{\partial^2 v}{\partial x^2} \qquad (8.11)$$

The call price is calculated from the Green function $v = g^\dagger(x, t|x_T, T)$ of this pde (where the dagger denotes the adjoint of g). The forward-time Kolmogorov pde

$$\frac{\partial g}{\partial T} = -\frac{\partial}{\partial x_T}((r - D(x_T, T)/2)g) + \frac{\partial^2}{\partial x_T^2}\left(\frac{D(x_T, T)}{2}g\right) \qquad (8.12)$$

has exactly the same Green function $g(x_T, T|x, t) = g^+(x, t|x_T, T)$. The price sde corresponding to this Fokker–Planck pde (dropping subscripts capital T for convenience) is

$$dp = rp dt + \sqrt{p^2 d(p, t)} dB \qquad (8.13)$$

where $d(p,t) = D(x,t)$ and r is the risk-neutral rate of return (actually, r is arbitrary in the delta hedge and can be chosen freely). With $y = x - rt$ and $g(x, t|x', t') = G(y, t|y', t')$ (since $dx = dy$) we obtain

$$\frac{\partial G}{\partial t} = -\frac{\partial}{\partial y}\left(-\frac{E}{2}G\right) + \frac{\partial^2}{\partial y^2}\left(\frac{E}{2}G\right) \qquad (8.14)$$

with $E(y,t) = D(x,t)$ which has the sde

$$dy = -E(y,t)dt/2 + \sqrt{E(y,t)}dB(t) \qquad (8.15)$$

and yields the corresponding price sde (with $x = \ln S(t)/S_c(t)$)

$$dS = \sqrt{S^2 e(S,t)} dB(t) \tag{8.16}$$

with price diffusion coefficient $e(S,t) = E(y,t)$. All of this shows that the risk-neutral discounted price $S = pe^{-rt}$ is a Martingale. However, the expected return on the stock appears in the consensus price in the diffusion coefficient. Unlike the Gaussian returns model, the stock return cannot be completely eliminated by constructing a Martingale.

8.3 Pricing options approximately via the exponential density

8.3.1 Normalization

In order that the exponential density

$$f_1(x,t) = \begin{cases} Ae^{\gamma(x-\delta)}, & x < \delta \\ Be^{-\nu(x-\delta)}, & x > \delta \end{cases} \tag{8.17}$$

with the slope jump location δ to be determined, satisfies the diffusion pde

$$\frac{\partial f_1}{\partial t} = -R(t)\frac{\partial f_1}{\partial x} + \frac{1}{2}\frac{\partial^2 (Df_1)}{\partial x^2} \tag{8.18}$$

it's necessary to remove the delta function at $x = \delta$ arising from the slope discontinuity. The solutions below lead to the conclusion that R is continuous across the discontinuity, and that $D(x,t)$ is discontinuous at $x = \delta$.

In order to satisfy conservation of probability at the discontinuity at $x = \delta$ it's not enough to match the current densities on both sides of the jump. Instead, we must apply the more general condition

$$\frac{d}{dt}\left(\int_{-\infty}^{\delta} f_-(x,t)dx + \int_{\delta}^{\infty} f_+(x,t)dx\right) = \left.\left((R - \dot{\delta})f - \frac{1}{2}(Df)'\right)\right|_{\delta} = 0 \tag{8.19}$$

The extra term arises from the fact that the limit of integration δ depends on the time. In differentiating the product Df while using

$$f(x,t) = \theta(x - \delta)f_+ + \theta(\delta - x)f_- \tag{8.20}$$

and

$$D(x,t) = \theta(x - \delta)D_+ + \theta(\delta - x)D_- \tag{8.21}$$

we obtain a delta function at $x = \delta$. The coefficient of the delta function vanishes if we choose

$$D_+ f_+ = D_- f_- \tag{8.22}$$

8.3 Pricing options approximately via the exponential density

at $x = \delta$. These conditions determine the normalization coefficients A and B once we know both pieces of the function D at $x = \delta$. In addition, there is the extra condition on δ,

$$(R - \dot{\delta})f\big|_\delta = 0 \tag{8.23}$$

so that $\delta = \int R(t)dt$. With

$$D_\pm(x,t) = \begin{cases} D_+(1 + v(x - \delta)), x > \delta \\ D_-(1 - \gamma(x - \delta)), x < \delta \end{cases} \tag{8.24}$$

we obtain

$$\frac{A}{\gamma^2} = \frac{B}{v^2} \tag{8.25}$$

From the normalization condition $\int f_1 dx = 1$ follows

$$\frac{A}{\gamma} + \frac{B}{v} = 1 \tag{8.26}$$

Combining (8.25) and (8.26) yields the normalization

$$A = \frac{\gamma^2}{\gamma + v}$$
$$B = \frac{v^2}{\gamma + v} \tag{8.27}$$

Finally, we see from (8.27) that scaling $f_1(x,t) = t^{-H}F(x/t^H)$ can hold if and only if γ and v have exactly the same dependence on t^H. That is, the slopes of $\ln(f_1)$ can differ to the right and left of δ but the time dependence must be the same on both sides of the discontinuity in order for scaling to hold.

With $R(t)$ continuous we would have a trivial time evolution of value. With $x = \ln(p(t)/p_c)$ with p_c a constant locating the peak of f_1 for $t = 0$ we see that $x - \delta = \ln(p(t)/p_c(t))$ where $p_c(t) = p_c e^{\int R(t)dt}$. This means that the stock interest rate $R(t)$ appears in the Martingale option price. Statistically seen, the expected stock interest rate $R(t)$ is an unreliable parameter. We can eliminate it by applying the working man's Martingale condition

$$\langle p(t) \rangle = p^{r_d(T-t)} \tag{8.28}$$

where p is the observed stock price at the present time t, and r_d is the cost of carry (the risk-free interest rate plus a few percent).

8.3.2 Exponential option pricing

We've seen that we can extract the one-point density from FX data only in the small intraday scaling region; that density is exponential with $H \approx 0.35$. Our best practical guess for interday trading is to assume an exponential density with $H \approx 1/2$.

Consider the price of a call for $x > \delta$ where $u = (x - \delta)/\sqrt{t}$, where $R(t)$ is determined by the working man's Martingale condition (8.28), yielding

$$r_d = \frac{1}{(T-t)} \left(\int_t^{t+T} R(s)\,ds + \ln\left(\frac{\gamma v + (v - \gamma)}{(\gamma + 1)(v - 1)} \right) \right) \tag{8.29}$$

which fixes R.

With

$$C(p, K, T - t) = e^{r(t-T)} \int_{\ln K/p}^{\infty} (p_T - K) g(x_T, T; x, t)\,dx_T \tag{8.30}$$

where p is the known stock price at present time $t < T$. We know the Green function both empirically and analytically only for the case where $g(x,t|0,0) = f_1(x,t)$. This approximation yields

$$C(K, p, \Delta t) = e^{-r_d \Delta t} \langle (p_T - K)\theta(p_T - K) \rangle$$
$$= e^{-r_d \Delta t} \int_{\ln(K/p)}^{\infty} (pe^x - K) f_1(x, t)\,dx \tag{8.31}$$

and amounts to assuming that $p \approx p_c$, that the present price is the same as the consensus price. We'll see that this uncontrolled approximation does not destroy the usefulness of the prediction. In addition, in agreement with traders, we've replaced the risk-free interest rate by the cost of carry rate in the prefactor in (8.31). If we also make the approximation $R(x,t) = r - D(x,t)/2 \approx R(t)$ then for $D(x,t)$ linear in $|x|/|t|^{1/2}$ we obtain the exponential density (8.17). We can take this as a phenomenological prediction of the option price.

Given this approximation, with the exponential density (8.17) and normalization (8.27), we find that the call price is given for $x_K = \ln(K/p) < \delta$ by

$$C(K, p, \Delta t)e^{r_d \Delta t} = \frac{pe^{R\Delta t}}{(\gamma + v)} \frac{\gamma^2(v - 1) + v^2(\gamma + 1)}{(\gamma + 1)(v - 1)} + \frac{K\gamma}{(\gamma + 1)(\gamma + v)} \left(\frac{K}{p}e^{-R\Delta t}\right)^{\gamma} - K \tag{8.32}$$

For $x_K > \delta$ the call price is given by

$$C(K, p, \Delta t)e^{r_d \Delta t} = \frac{K}{\gamma + v} \frac{v}{v - 1} \left(\frac{K}{p}e^{-R\Delta t}\right)^{-v} \tag{8.33}$$

The corresponding put prices are

$$P(K,p,\Delta t)e^{r_d \Delta t} = \frac{K\gamma}{(\gamma+\nu)(\gamma+1)}\left(\frac{K}{p}e^{-R\Delta t}\right)^{\gamma} \tag{8.34}$$

if $x_K < \delta$ and

$$P(K,p,\Delta t)e^{r_d \Delta t} = K - \frac{pe^{rR\Delta t}}{(\gamma+\nu)}\frac{\gamma^2(\nu-1)+\nu^2(\gamma+1)}{(\gamma+1)(\nu-1)}$$
$$+ \frac{K\nu}{(\nu+\gamma)(\nu-1)}\left(\frac{K}{p}e^{R\Delta t}\right)^{-\nu} \tag{8.35}$$

for $x_K > \delta$. These predictions were first derived c. 1990 by Gunaratne, who accidentally introduced the notation $\gamma\nu$.

8.4 Option pricing with fat tails

Consider the price of a call for $x > \delta$,

$$C(p,K,T-t) = e^{r(t-T)} \int_{\ln K/p}^{\infty} (p_T - K)g(x_T,T|x,t)dx_T \tag{8.36}$$

We know the transition density analytically only for the case where $g(x,t|0,0) = f_1(x,t)$, the empirical distribution for the case where we can make the approximation $R(x,t) = r - D(x,t)/2 \approx R(t)$,

$$C(p_c,K,T-t) = e^{r(t-T)} \int_{\ln K/p_c}^{\infty} (p_T - K)f(x_T,T)dx_T \tag{8.37}$$

This is enough to make our point: with fat tails $f_1(x,t) \approx |x|^{-\mu}$, $|x| \gg 1$ we get

$$C(p_c,K,T-t) \approx e^{r(t-T)} \int_{\ln K/p_c}^{\infty} pe^x x^{-\mu} dx = \infty \tag{8.38}$$

Fat tails cause the option price to diverge. We haven't found fat tails in intraday FX data; the densities generally cannot be extracted even for a time scale of a single day. So we don't know if fat tails are present in stock data. They may well be. But if they are, then they're apparently ignored by traders pricing options.

8.5 Portfolio insurance and the 1987 crash

Option pricing is based on the assumption of a normal liquid market, a "Brownian" market, otherwise there is no way to price options. This assumption fails miserably when liquidity dries up in a market crash.

Synthetic options are based on put-call parity (Chapter 5). Under normal market conditions, a synthetic option duplicates the payoff of a long underlying position with a long call and short put at the same strike and expiration. For example, a synthetic call is constructed by the right-hand side of $C = P + V - e^{-r_0(T-t)}K$, and money in the bank would hypothetically be equivalent to $e^{-r_0(T-t)}K = P + V - C$. Anyone who takes this literally should also believe in the tooth fairy, but adults who do not believe at all in the tooth fairy believed in synthetic options because they ignored liquidity droughts in market crashes. As Morris (2008) wrote, only people of high intelligence can make monumental mistakes.

Portfolio insurance, as engineered by Rubenstein and Leland (1981), is based on synthetic options. The assumption is that one should go with the market: buy as prices rise above some benchmark price increase, and sell when prices fall below some benchmark drop. This is the destabilizing behavior typical of our modern era. When enough big traders act collectively, when their computer programs try to execute the same sell orders massively, a large enough drop in market prices can cause a crash. That is presumably what happened in 1987, when the New York stock market fell by 40% in a few days.

8.6 Collateralized mortgage obligations

Collateralized mortgage obligations (CMOs) were invented in the 1980s, exactly in the era when finance market deregulation led to the inability of savings and loans to compete further in mortgage lending (Lewis, 1989). With a CMO, mortgages are split into three separate derivatives and sold separately (Morris, 2008; Soros, 2008). The construction of special derivatives made the parts look superficially like bonds. With too much money in circulation looking for a place to be parked at high interest rates (see Chapter 9 for the reason), Wall Street bought mortgages and resold them at a profit as CMOs. As with the derivatives market in general, complex financial instruments were created that no one understood. In particular, the derivatives could not be evaluated in any sensible way if liquidity were to start drying up. The role played by CMOs in the 2007 subprime mortgage fiasco, a part of the larger worldwide credit bubble, is described by Morris (2008) and is discussed in the context of unregulated forms of money creation in Chapter 9.

We've formulated this chapter in standard textbook fashion, assuming that a local currency can be treated as risk-free. The question of the time scale over which that assumption makes sense was not considered. The next chapter is devoted to the history of the instability of the Dollar, and in the few months over which this chapter was written the price of oil increased from about $80/barrel to $140/barrel. Normally, one assumes that a major currency can be taken to be approximately risk-free on a time scale of weeks or even months. On a time scale of the last few months, the Dollar was not very reliable in the role of a "risk-free" paper asset. We will next present the case that we live in a very singular era in financial history, the era of the worldwide credit bubble based on a weak and further weakening Dollar as international reserve currency.

Credit default swaps have also been modeled and used by lending agencies in deciding on the riskiness of loans. What the modelers and users typically and too easily forget, or were never aware of, is that all such models assume fairly normal market liquidity; the models cannot be used to anticipate a liquidity drought, a market crash.

9
FX market globalization
Evolution of the Dollar to worldwide reserve currency

9.1 Introduction

We return to the theme introduced in the first chapter, economists' expectations of stable equilibrium vs the reality of market instability under deregulation. We'll now illustrate the disparity by following the evolution of the Dollar and FX markets from the gold standard to September, 2008.

We begin by following Eichengreen's (1996) informative history of the evolution of western FX markets from the gold standard of the late nineteenth century through the Bretton Woods Agreement (post-WWII–1971) and later the floating currencies of the early market deregulation era 1971–1995. We add equations, models, and observations to broaden that discussion and explain mathematically how the FX markets work. We also add a discussion of the era 1995–2008 based on our understanding of finance markets gained in Chapter 7.

Although WWI-era finance data (or any pre-computerization-era data) would be too sparse to permit a meaningful empirical analysis (see Chapter 7), there is qualitative evidence for a change from stability to instability over the time interval of WWI. With the risk and instability of our present era (Chapters 7 and 8) in mind, we show how speculators could have made money systematically from an *effectively regulated* FX market like that of the gold standard era. The present era normal liquid FX markets are in contrast approximately impossible to beat, are "efficient," and require options to hedge against currency risk. The ideas of Martingales and options/hedging became of great practical importance after 1971, but were of little use or interest in the gold standard era. The main weakness of Eichengreen's book is that the role of derivatives as a completely unregulated form of money creation is ignored. Before following the evolution of the Dollar, we first define the "official" or "on the books" money supply

M0–M3, and the regulated method of money creation called "fractional reserve banking."

9.2 The money supply and nonconservation of money

Standard measures of the money supply, money recorded "on the central bank books," are defined here: M0 is the total of all "physical currency," plus accounts at the central bank that can be exchanged for physical currency. That is, M0 consists of physical currency circulating in the economy plus checking account deposits. This is a measure used by economists trying to quantify the amount of money in circulation, and is a very liquid measure of the money supply. M1 consists of currency in circulation plus checking deposits, officially called demand deposits, and other deposits that work like checking deposits, plus traveler's checks. M2 consists of M1 plus most savings accounts, money market accounts, and small-denomination time deposits (certificates of deposit of under $100,000). M2 provides us with the measure of the currency within the country. M3 is extremely important, as compared with M2: M3 includes M2 plus all other CDs, deposits of Eurodollars, and repurchase agreements ("repos"). "Eurodollars" means simply Dollars in foreign banks anywhere in the world, and can be used to create Dollar-denominated credit in foreign banks via fractional reserve banking. Eurodollars are beyond the control of the US Federal Reserve Bank. M3/M2 tells us the fraction of the currency beyond the control of the central bank that is supposed to regulate that currency.

Fractional reserve banking is a form of money creation via credit (http://en.wikipedia.org/wiki/Money_creation) regulated by the central bank. As an example, consider a saver with $100, a bank accepting a $100 deposit, an outboard store and a customer who buys from the store using only borrowed money. If the fractional reserve rate is 20%, then from $100 deposited the bank must keep on hand $20 and may lend $80 to the consumer. The consumer borrows $80 and buys some motor parts. The dealer deposits that $80 in the bank, of which $64 may then be lent to the boater who wants to buy several cases of oil. The dealer deposits that $64 in the bank, and (if the bank is willing to give the loan) the boater may borrow another $51.20, and so on. Money is effectively *not* conserved because the time scale for borrowing is very short compared with the time scale for repayment (consider credit cards, for example, where the repayment time may be set approximately equal to infinity). Furthermore, if there is default then the money is never repaid; it was created and remains in circulation. This is the regulated form of money creation under the legal banking system. A central bank's method of trying to control the money supply is simple: selling treasury bonds and treasury bills to banks reduces the money supply, while buying them back from banks increases the money supply. Selling US Treasury Bills and bonds to foreigners is the government's way of borrowing money from foreigners.

9.3 The gold standard

Credit has played a strong role in finance and economics since at least the Renaissance in Europe, and dominates all markets today. With a credit card, purchases can be made with the tap of a computer key. Let's begin next with the pre-WWI era of financial stability when the level of money creation via credit was relatively low.

The gold standard became widely accepted around 1890. But even then the "quantity theory of money" presented by Hume (1898) did not hold strictly. With a strict gold standard, and no new mining production or coinage of existing gold, money would be conserved. We've explained how money is created via credit under fractional reserve banking. Credit controlled by central banks is included systematically in the estimate of both the national (M2) and worldwide (M3) money supply.

Before WWI, stable currency values were supposedly maintained by the threat of central bank intervention. The threat led speculators to bid up a weak currency with the expectation of a profit, and thereby strengthened the currency via a self-fulfilling process: speculation in that era tended to stabilize FX markets, as we'll explain below via a specific model. After WWI the central bank threat either fell by the wayside or else no longer carried sufficient weight, and FX markets became unstable: weak currencies were bid lower by speculators. The historic reasons for the change are discussed below.

9.4 How FX market stability worked on the gold standard

Adhering to a gold standard meant very tight money: $20 would buy you an ounce of gold, but money was hard to come by. Credit cards didn't exist; credit was hard to obtain. Banks in the gold standard era, and later, did not make loans for consumption. Today, high school students are offered credit cards, but in the 1960s a college student could not borrow $250 from a bank to buy an old car unless his parents co-signed the loan. Americans who came of age after 1971 have grown up in a very different country, one that since 1981 seems very different to me. Markets were relatively illiquid, meaning that items were infrequently traded. We begin with the pre-WWI era, in which maintaining the gold value of a currency was the overwhelmingly dominant factor in finance. According to the history of the gold standard and its replacement as international standard by the post-WWII Dollar under the Bretton Woods Agreement, we can infer that there was a fundamental shift in the FX noise distribution after WWI from a stationary process to a Martingale process. The shift was from a stationary market to a nonstationary one.

The motion of money in a political economy bears only very limited resemblance to the motion of a mass in physics. In economics/politics, the future can be shaped by acting on beliefs and wishes. The beliefs/wishes are then called self-fulfilling expectations. Changing the motion by acting on wishes is impossible in the mindless matter that we study in physics. Therefore, we must pay close attention to political policy in discussing economics and finance, where the log return "$x(t)$" bears only a limited resemblance to $x(t)$ in physics. A main question in *political economy* is: what should be regulated, and what should be allowed to move freely? For example, telephone costs have dropped significantly under deregulation (phone calls to and from Europe are possible at about $0.03/minute now in either direction) but electricity costs for consumers have not decreased under deregulation, nor has service become better. Information transport is very cheap; electricity transport is extremely expensive. The question of deregulation is nontrivial. In particular, we will analyze whether money creation and financial transfers define a self-stabilizing dynamical system, or do we need regulations to achieve market stability?

In all that follows we must keep in mind two different degrees of instability. First, in a normal liquid market instability means simply that the market returns are a nonstationary process: statistical equilibrium is not approached. This is modeled in Chapter 7, and examples are provided by the usual daily operation of a finance market. Second, a market crash, a liquidity drought, is a far worse form of instability that we cannot model reliably because no meaningful empirical analysis is possible (the statistics would be too sparse; see Chapter 7). An example from 2007 is the subprime mortgage fiasco, and consequent related examples from 2008 are the bankruptcies of Lehman Brothers and AIG, threatening either the collapse of the worldwide financial system (depression) or the further inflation of the Dollar (*why* these are the alternatives is discussed in this chapter). Laws were passed in the USA in the 1930s to avoid and manage liquidity droughts, but most of those laws have been repealed since the deregulation revolution of 1981. So let's turn to history as a guide for our analysis.

Before WWI, the main job of western central banks and parliaments was seen as keeping the national currency from falling outside chosen gold standard "bands." The claim is that, before WWI, the currency speculators, confident that governments could be relied on to maintain the gold value of the currency, bid a weak currency up, expecting a profit in the long run. The older FX data are too poor to test this idea but, if true, it would mean that FX markets were (at least asymptotically) stationary in that era. This is interesting because no known market is stationary today. We know that FX markets since 1958 are described approximately by the nonstationary Gaussian returns model. Gaussian returns markets are nonvolatile; volatile markets became important after the crash of 1987. Although economists generally do not recognize market instability as a key idea, or as *any* admissible idea, the economist Eichengreen (1996) argues that the onset of the instability coincided with social pressures after WWI (his subsequent papers are all based on equilibrium models, however).

Here's how the pre-WWI FX market worked. Imagine a Dollar equivalent to 25.9 grains of gold. Take the Reichsmark (the "Euro" of that time) as the foreign currency of interest, and focus on trade with Germany. Assume that credit (money creation without gold backing) doesn't change the money supply significantly inside the US. A trade deficit meant too many Dollars were outside the country. In practical terms, M3/M2 was too large. When there were too few Dollars inside the country, economic activity within the US fell. Banks in that era attempted in some rough sense to conserve the Dollar, so that the trade deficit reduced liquidity inside a country on the gold standard (meaning deflation, lower prices, unless more money was printed). So the trade deficit was eventually reversed via cheaper exports. The latter brought money back into the country, which increased the Dollar against the RM without the need for a devaluation of the weak Dollar by the central bank. By reducing the money supply (thus weakening demand further), a central bank could speed up this process.

We can make an FX model describing that stability. Consider the logarithmic return $x(t) = \ln(p(t)/p_c)$ where p is the price of one currency in units of another (e.g., the Reichsmark in 1913, or the Euro today, in Dollars), and p_c is the value of the Dollar set by the gold standard. In a *stationary* process, the one-point returns density $f_1(x,t)$ is time-independent: the average return, the variance, and all other moments of the one-point distribution are constants. A market that possesses a statistical equilibrium distribution has fluctuations that obey a stationary process $x(t)$. We can easily model an asymptotically stationary market. From the usual stochastic supply–demand equation

$$dp = rp dt + \sigma_1 p dB(t) \tag{9.1}$$

we obtain

$$dx = (r - \sigma_1^2/2)dt + \sigma_1 dB(t) \qquad (9.2)$$

Let $R = r - \sigma_1^2/2$ denote the expected return, $x = Rt$. For FX markets we know empirically that $R \approx 0$. If $-\infty < x < \infty$, then (9.2) is the lognormal model introduced by Osborne in 1958 and used by Black and Scholes in 1973. The speculators' behavior generates the noise, which in the case of (9.2) is the Wiener process. But central bank intervention means that unbounded prices, $-\infty < x < \infty$, is the wrong assumption.

With the pre-WWI Dollar supported within a gold band $b_1 < x < b_2$, stationarity is the consequence. We can set the equilibrium probability density $f_1(x) =$ constant except at the boundaries and then we obtain an approach to statistical equilibrium: $f_1(x,t)$ approaches $f_1(x)$ as t increases (see Stratonovich (1963) for the mathematical details). That is, the market is asymptotically stationary. Here's how speculators could systematically suck money out of a stationary market. Consider the price distribution $g(p,t) = f(x,t)dx/dp$ with price variance σ_p^2. One buys, for example, if $p < p_G - \sigma_p$, and one sells if $p > p_G + \sigma_p$. Such fluctuations are guaranteed if the stationary process is *ergodic* or is at least recurrent (discrete stationary processes are recurrent (Kac, 1949)), and the first passage time for a specific fluctuation can easily be calculated (see Stratonovich (1963) or Durrett (1984; 1996)). So we understand how speculators could systematically have made money *with little risk* in the pre-WWI era. All that was required was, once the bet was placed, the trader had to be able to afford to leave his money in place until stationarity provided him with a gain. That is, the correct strategy was to buy and hold; there was little or no motivation to hedge risk, or to trade frequently. The stabilization process was a self-fulfilling expectation. But there's a very good reason why speculators traded in a stabilizing way: they were limited by boundaries imposed by the central bank.

We must interpret the boundary conditions in order to understand why traders stabilized the FX rate: it wasn't the gold standard alone, but rather was the serious threat of punishment combined with reward that produced stability. The band limits, $b_1 < x < b_2$, represent the threat of intervention and can be understood effectively as a form of regulation. The process is asymptotically stationary if and only if both b_1 and b_2 are finite (the particle with position x is confined between two walls), so that $\langle x \rangle = (b_1 + b_2)/2 =$ constant fixes p_c at the gold value of the Dollar. The central bank would threaten to intervene to buy/sell Dollars if x would hit b_1 or b_2, so speculators could confidently buy Dollars if $\sigma < x < b_1$, for example, where $\sigma^2 = \langle x^2 \rangle =$ constant. Stationarity guarantees that profitable fluctuations occur with average first passage time $\tau = \sigma^2/2\sigma_1^2$. The ability to establish boundaries, representing "threats of punishment with teeth," a form of regulation, generated stable dynamics. We can profitably emphasize the stabilization process via comparison with related social analogies.

In the language of the Ultimatum Game, the boundary conditions/regulations were threats of financial punishment. Those threats of punishment were effective if the central banks had gold reserves large enough to beat the speculators, if necessary. A related social analogy is the old saying that "kids like boundaries." Kids don't "like" boundaries, everyone prefers to be free to explore vast possibilities, including the dangerous ones like sex, alcohol, and habitual smoking at ages 14–17, but if parents set strong boundaries and enforce them with punishment and rewards then behavior modification is likely. This provides us with an example of a regulated system like the FX market on the gold standard. Here's an example of an unregulated free market, "The Tragedy of the Commons (*die Tragödie der Allmende*)": with free farmers sharing a common meadow, the tendency is for each farmer to add "one more cow." Regarding stability or lack of it, Adam Smith wrote earlier that moral restraint is required for a free market system to function. Moral restraint doesn't prevent farmers from adding one more cow. Moral restraint is inadequate in modern finance. A comparison of the gold standard era FX market with current FX markets illuminates this claim. By "morals," should we include the notion of not taking advantage of a big personal gain if other citizens would be significantly hurt by that gain?

Summarizing, speculators created stabilizing self-fulfilling prophecies before WWI because governments (a) had adequate gold reserves and (b) saw their job as maintaining the stability of the currency, instead of guaranteeing high employment and social services. WWI changed the priorities. The rise of socialism and labor unions after WWI meant that social spending had to be given priority. The consequent threat of inflation via printing paper money or borrowing to finance deficits caused the wealthy to prefer gold over paper Dollars. Social spending and regulation on finance markets increased dramatically during and after the Great Depression of the 1930s.

9.5 FX markets from WWI to WWII

Banks and the government saw the avoidance of inflation, not high employment, as their main job prior to the depression. After the onset of depression, Keynesian policies were introduced in order to fight it. Keynesian policies were seen as inflationary because fiscal policies should be instituted to stimulate spending. That is, the money supply should be inflated, distributing money to those in need who would spend it for necessities.

From an entirely different perspective, inflation of the money supply is necessary but not sufficient for a boom or bubble (Kindleberger, 1996), and inflation of the money supply via credit played a role in creating and expanding the 1929 bubble. The stock market crash of 1929 caused a liquidity crisis:[1] deflation occurred because there was no "lender of the last resort" to provide liquidity. Margin trading, a form of leveraged betting, was a major source of running up stock prices in the late 1920s, as in the late 1990s. Therefore, the 1929 crash hurt the lenders, including banks that had lent money for stock speculation (this can be said of the 2007–08 crisis as well). Many banks went bankrupt and closed after the crash. Depositors withdrew money from solvent banks for fear of losing more money, causing a liquidity drought. The depression/deflation followed from the lack of money in circulation: many people were unemployed, and those with money tended to hoard instead of spending. *Bank deposits were not insured at that time.* Franklin D. Roosevelt was elected President in 1932 based on his promise to abandon past social policies and institute a "back to work" policy, and to restore confidence in the banking system. People with money expected inflation via social spending, encouraging the conversion of paper Dollars into gold at $20/ounce, a conversion rate that reflected the financial stability of a dead era. The liquidity crisis can be understood as a form of the Gambler's Ruin (see Chapter 4). Until 1935, the gold value of the Dollar had been maintained roughly at $20/ounce (from a physical perspective, this was possible because the US gold supply had been analogous to a heat reservoir).

Roosevelt's "Bank Holiday" in 1933 was partly the consequence of a run on gold by people getting rid of the Dollar. To close escape hatches, in 1935 he outlawed the ownership of gold by Americans, recalled all gold coins, excepting rare coins held by collectors, and then fixed the price of gold at $35/ounce (potentially inflating the money supply by about a factor of two!) thereby guaranteeing that Americans could not depreciate the Dollar by buying gold. Fear of bank failures was exceedingly widespread: many people hid money at home rather than trust the banks again. In a further effort to control the Dollar and restore confidence, bank deposits were insured and the Glass–Steagall Act was passed to keep commercial banks out of the stock brokerage business. With New Deal acts in place as law, the US government could then spend freely on public works projects like the Tennessee Valley

---

[1] The key question for the reader is: why haven't market crashes since 1929 but before 2007 caused another depression?

Authority, which was and still is a very effective and profitable socialist project, and the Works Projects Administration. The US government got away with it because the US then had the largest gold reserves in the world. France and, surprisingly, the defeated Germany were second and third.[2] During the depression, the gold reserves of the US, France, and Germany were more than adequate compared with the paper currency in circulation. Liquidity, not total wealth, was the problem.

9.6 The era of "adjustable pegged" FX rates

After WWII, eastern Europe became Soviet satellite states while Germany under the Marshall Plan was rebuilt as a capitalist showcase against the USSR. This caused large Dollar transfers to Europe. The gold standard was replaced in the west by the Bretton Woods Agreement: (1) "adjustable" pegged exchange rates, (2) controls were allowed and placed on international capital flows, and (3) the IMF was created to monitor economic policies within participating nations, and to extend credit to countries at risk with large trade imbalances.[3] Interest rates were capped as well. Controls were understood as necessary in order to avoid flight from a currency; the commitment to economic growth and/or full employment is not consistent with absence of inflation. Financial markets in that era were clearly regulated.

By 1959, the Bretton Woods exchange controls began to fail. The US had pegged the Dollar artificially to internally nonliquid gold while inflating the Dollar, and the rest of the west pegged currencies to the Dollar with the right to exchange Dollars for gold. The Dollar had replaced gold as the unit of international currency. Japan and Germany restricted US imports further. The result was that, by 1959–60, the order of magnitude of Dollars in Europe was on the order of magnitude of the Dollar value of gold stored in Fort Knox (the latter was about $20,000,000). *We don't need a detailed dynamical model to help us to understand that speculators rightfully expected a devaluation of the Dollar.* In 1958, Eisenhower prohibited Americans from owning gold in Europe, and in 1961, Kennedy outlawed the ownership of gold coins by American collectors, going further than Roosevelt, but those acts were like band aids on a broken artery. The acts were a futile attempt to delay the

---

[2] A common misunderstanding is that the Nazis came to power in 1933 because the Allies drained Germany via reparations. In fact, Germany successfully resisted reparations but had high unemployment for the same reason as did the USA. Hitler's *Finanzminister* Hjalmar Horace Greely Schacht got the ball rolling via Keynesian-style inflationary public spending. That was the era when the Autobahns began to be built, for example. The wild German inflation of 1923 was a Berlin ploy to deflect France's demands for reparations payments.

[3] The traditional method of removing a serious trade imbalance was devaluation of the currency against gold.

inevitable: either transfer all the gold in Fort Knox on demand to the creditors, or else devalue the Dollar. Significantly for the history of the Dollar, the Organization of the Petroleum Exporting Countries (OPEC) was formed in the same era.

We can model the post-WWI instability mathematically. Far-from-equilibrium dynamics is exhibited by the Gaussian returns model that Osborne proposed empirically to describe stock returns in 1958,

$$dx = (r - \sigma_1^2/2)dt + \sigma_1 dB(t) \qquad (9.3)$$

subject to no boundary conditions. Even with $r < 0$ this model allows no approach to statistical equilibrium. The model is nonstationary but nonvolatile. Presumably, FX markets were like this from the end of WWI through 1987. Volatility could be introduced in an artificial way by letting the constant σ_1 experience large jumps at discrete, unpredictable times.

Modern credit based on fractional reserve banking is a regulated form of money creation. The first credit card, the BankAmericard (later VISA) appeared c. 1960, introduced by an Italian-American banker in California presumably to help the local Italian community. By 1964 the Dollar was weak enough that silver coins were worth more than their face value, so the USA under Lyndon Johnson stopped minting them. The Vietnam War caused inflation, as do all wars.[4] The government-financed Advanced Research Projects Agency Network (ARPANET) appeared in that era, signaling that communication speed would increase in the future. These facts are central for understanding today's Dollar, and the FX transfers that occur second by second, shortening the time scale over which instabilities can make their effects known.

9.7 Emergence of deregulation

In 1961 the order of magnitude of Dollars outside the USA was far greater than the US gold supply in Dollars. In 1971 France demanded an exchange of Dollars for gold. This forced Richard Nixon to deregulate the Dollar, to let it "float" freely against gold and all other currencies. Note from Figure 9.1 that M3 was still insignificant compared with M2, which means that the number of Dollars in circulation in the US was far greater than the gold supply at \$35/ounce. Significantly, Forex was created in 1971, and the first ARPANET email program appeared then as well. The Chicago

---

[4] No well-off population would likely choose war were the costs explained in advance, and were it understood that the war must eventually be paid for via either higher taxes, inflation, or generally both.

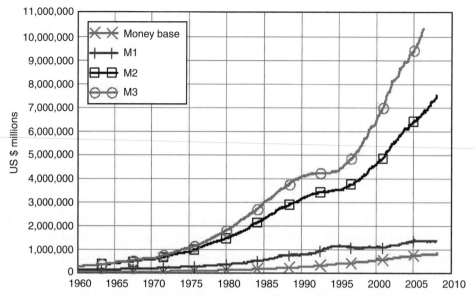

Figure 9.1 The US Dollar money supply, showing especially the growth of M3, reflecting the Dollar as the world's default reserve currency since WWII (provided by Steve Keen's "Debtwatch," www.debtdeflation.com/blogs). The standard measures of the money supply are: M0 = the total of all physical currency, plus accounts at the central bank that can be exchanged for physical currency. M1 = M0 + those portions of M0 held as reserves or vault cash + the amount in demand accounts ("checking" or "current" accounts). M2 = M1 + most savings accounts, money market accounts, and small denomination time deposits (certificates of deposit of under \$100,000). M3 = M2 + all other CDs, deposits of Eurodollars and repurchase agreements. Eurodollars = Dollars in foreign banks, and can be used to create Dollar-denominated credit in those banks beyond control of the US Federal Reserve Bank.

Board of Options Exchange (CBOE) was created in 1973, the year that the Black–Scholes solution was finally published. This was the beginning of deregulation and the consequent financial revolution of 1971. The creation of derivatives/options/hedging literally exploded, and we understand why: the new market instability of the Dollar required hedging bets.

OPEC understood the 1971 Dollar devaluation (deregulation from gold, basically defaulting on payments to Europeans holding Dollars) and raised the price of oil dramatically. By 1973 gold had hit \$800/ounce, options trading was in full swing, and many OPEC oil fields were nationalized. To give the reader an idea of the price inflation, a VW Beetle that had cost \$700–800 in 1968 cost \$1600 by 1974, and the US gasoline price had doubled in Dollars as well, after first hitting \$1.20/gallon. I was there, waiting in the first gas lines in Houston. This was the great inflation of the 1970s, and was the beginning of deregulation in American politics. America from 1935 to 1971 was very, very different from America after 1971. The regulations put in place in the Great Depression (and irrationally hated as "evil" by the right wing) were under systematic attack by political and economic conservatives, who pointed to stagflation and announced: "Keynesianism doesn't work!". The program of privatization and deregulation began euphorically and in earnest, and the idea of making money by trading instead of producing began to spread broad and deep roots.

With the Dollar inflating faster than paychecks could keep up, Americans in the 1970s got rid of Dollars in favor of art, collectible coins, and other "values," and began to run up credit card bills with the expectation that next year's Dollar would be worth much less than today's (this was the beginning of modern consumerism). In the 1980s, the savings and loan associations were deregulated and went bankrupt, unable to compete with bond trading houses like Salomon Brothers who split principle and interest into separate derivatives for sale to the public (see former bond trader Lewis's excellent description (1989)). Leveraged buyouts financed by junk bonds[5] emerged, and became the order of the day on Wall Street. Old companies were bought and stripped for profit, with the carcass either later sold for profit or abandoned. Monetization of debt increased, with leveraged borrowing providing the required liquidity. FX transactions, on the order of $\$10^8$/minute c. 1981, decoupled from economic growth. With no certainty about currency at home or in international trade, physicists began to be hired on Wall Street to model derivatives in the 1980s. See Derman (2004) for an excellent and entertaining history of that era. The October, 1957 launching of Sputnik had caused the USA to pour money into science education in the 1960s, and by 1971 there was an oversupply of physics graduates available to be hired elsewhere, including as "rocket scientists" on Wall Street. Emanuel Derman was one of the first.

The inflation rate became so high (the USA experienced "stagflation," meaning inflation combined with unemployment) that Keynesian economics fell into disrepute, and monetarists claimed victory. Neo-classical economics theory was revived and began to fill the academic void (created by the fall of Keynesian economics) under the heading of "rational expectations." As my good friend and economist Duncan Foley once stated, you had to live through that era in order to understand the magnitude of the change in the

---

[5] A junk bond fund operates on the assumption that the danger of individual risky bonds is reduced by choosing a basket of risky bonds.

(academic economics) profession. Even liberals became free market theorists. The completely nonempirically based model called "rational expectations" provides the theoretical underpinning for unlimited deregulation and privatization.

At street level, to fight inflation, Carter-appointed Federal Reserve Bank Chairman Paul Volcker let interest rates float to 13.5% in 1981, eventually reducing inflation to 3.5% by 1983. The budget-breaking by the US government was far worse in the eras 1981–88 and 2001–08 than in any other era. In both eight-year terms, two presidents were elected who had promised to reduce both the size of government and taxes simultaneously. Instead, the government budget deficit was vastly *expanded* while taxes were *decreased*. Adhering to the philosophy of *laissez faire*, they even encouraged US manufacturing capacity to leave the country in search of cheap labor costs (Mexico, China, etc.). The dominant idea of morals was to permit a few people to make as much money as possible. That philosophy was most forcefully expressed by the wealthy (by inheritance, not work) political pundit William F. Buckley, beginning with his trumped-up, scurrilous attacks on Yale professors in the 1950s (Yale Alumni Magazine, 2008).

Developments in technology have also played a central role in the financial revolution. Faster and cheaper communication encourages faster and greater financial transactions. By 1985 Apple (and Commodore) computers and PCs had become common in offices and homes. Discount brokers appeared as a form of finance market deregulation. In the 1970s and earlier, unless one had enough money to bring the buy/sell rates for stocks down to 3% or lower, one phoned one's stock broker and paid 6% coming and going, and orders were executed at a snail's pace. Limit orders were not effective because of the slowness of execution, unless you were a big enough player to have a trader on the exchange floor. By 1999 one could place a small (say $10,000–30,000 or a smaller amount) limit order with a discount broker on a Mac or PC and have it executed in a matter of seconds, if the order were placed close enough to the respective bid/ask prices (also shown in real time), paying $20 to buy and sell. Liquidity and speed of transaction increased by many orders of magnitude.

China had begun to invite foreign capital for building modern industry c. 1980, following the "opening of the door" by the Nixon–Kissinger visit with Mao and Zhou Enlai in the early seventies.[6] US industry had begun drifting across the Mexican border as early as 1960, and migrated later to

---

[6] The University of Houston began accepting large numbers of very good Chinese graduate students in 1981, as the result of an effort by Bambi Hu and myself.

9.7 Emergence of deregulation

Asia. The internal collapse of the USSR in 1991 signaled that capitalism had won, and that globalization via privatization and deregulation could begin in earnest. All of these facts are central for understanding the financial crisis of 2007–2008 (which will likely continue into 2009 and beyond) and the consequent return in 2008 of finance market regulation. The explosion of Dollar credit drove all of this.

In 1994, an extremist free market Congress was elected in the US, and the stock bubble inflated rapidly in response to the deregulation fever (Newt Gingrich and Tom DeLay led the charge in congress): *the Dow Jones Average quadrupled from 1994 to 2000. From 1987 to 1994 it had doubled, and had doubled earlier from 1973 to 1987.* By 1999 the signs of the bubble were strong: many people quit their jobs to "momentum trade" dot.com stocks on margin, others mortgaged their houses to bet on dot.com stocks that had never shown a profit (many never showed a profit). The dot.com bubble had begun with Netscape in August, 1995, and lasted through 2000–01. The World Wide Web was the result of 30 years of government development and investment, primarily for military purposes, and exploded financially when deregulated in 1995. No private business would have financed a non-profit development for 30 years.

Bubbles require credit for their inflation, and interest rate increases then create margin calls. The stock bubble was popped the same way as in 1929: the Fed tried to deflate it slowly via a systematic sequence of many small quarterly interest rate increases. The difference with 1929 is that, with an enormously inflated money supply (Figure 9.1) and lenders of the last resort in place to avoid a liquidity crunch, the air never completely came out of the market. With so many Dollars in circulation worldwide, investors, always rightly afraid of inflation and looking for gain, demand a place to park their money. Consequently, even in 2008 stocks sold at very high price/earnings ratios. As we've shown in earlier chapters, instability in a normal liquid market means that the ideas of "overvalued" and "undervalued" are effectively subjective so that stocks are "valued" at whatever the largest group of traders thinks they're worth at any given time. Valuation in a nonstationary market is largely subjective, and can shift like dunes in a hurricane.

In 1998, the world finance market nearly crashed again (Dunbar, 2000). Following two ideas, the M & M theorem and the expectation that market equilibrium will prevail after large deviations, the hedge fund LTCM had achieved a debt-to-equity ratio "approaching infinity," with leveraging supplied by nearly every major bank in the world. The fund was run by bond trader John Meriwether and economics Nobel Prize winners Scholes and Merton (Black had died in 1993; the Nobel Prize for the Black–Scholes model was awarded in 1997). Ignoring that markets are nonstationary and that the liquidity bath is necessary for the application of stochastic models in the first place, LTCM literally "became the market" in Russian bonds, so that when they wanted to sell they suffered the Gambler's Ruin. M & M had restricted their analysis to small changes in returns, where there is no need to worry about the Gambler's Ruin, and concluded that the debt-to-equity ratio doesn't matter in evaluating a firm. In contrast with their analysis, reality shows that the debt-to-equity ratio matters: too much debt leads to bankruptcy and can effectively eliminate the equity altogether. Since 2001 the US government began to operate with a dangerously high debt-to-equity ratio (see the next section). For a survey of finance market instabilities from the savings and loan days through LTCM and including the subprime mortgage fiasco, see Morris (2008). Unmanageable liquidity droughts are made possible today because finance markets in general, and derivatives markets in particular, have not been regulated (see also www.marketoracle.co.uk/Article3652.html).

Instead of gold era central bank "carrots and sticks" for speculators, we've relied on the IMF and other nondemocratic, supra-governmental agencies like the World Bank and the World Trade Organization that try to penalize participating governments who violate what amount largely to neo-classical-based playing rules. One idea is that governments should pay attention to maintaining "stable" exchange rates, but without any notion of limiting the highly leveraged derivatives trading that could easily bring down the global financial system (as in 1998 and 2008). Worldwide, a currency remains strong when an economically strong enough country has a decent trade balance and/or pays a high enough (compared with main competitors for deposits) interest rate. With a weak currency like the 2001–2008 (and beyond) Dollar, and with both trade and budget deficits out of control, the US government can finance its debt only through attracting foreign money via high enough interest rates or other waning influence. A small country like Argentina cannot get away with that sort of flagrant behavior. The USA has got away with fiscally irresponsible behavior from 1981 to 2008 only because speculators have not (yet) believed that the US Government will default on its financial obligations, but a large inflation is a kind of default. In analogy with the necessity to deregulate the Dollar from gold in 1971, we can expect a liquidity crisis whenever the bets in finance markets are on the order of magnitude of the money supply.

For perspective, the number of Dollars recorded officially in the money supply (M0–M3) in the world increased about 55% from 1945 to 1965, and by about 2000% from 1971 to 2001. "Money" includes credit. Governments

9.7 Emergence of deregulation

in our era have used interest rates to try to keep credit partially in check, also in Europe. For reference, when I first went to Germany in 1985 one could not buy a car on credit, and credit cards were useless in restaurants and gasoline stations (Scandinavia was an exception). At that time, credit cards were accepted in tourist businesses in big cities. Germans had a strong habit of saving and paying as they went, as did Americans before 1971. I recall vividly when, c. 1992, Deutsche Bank announced that credit cards would be issued and their use would be encouraged (the reader is invited to look up graphs of the growth of M2 and M3 for the Euro). When my wife and I hiked over the Alps in 1989, we sometimes had to take a bus to a larger Italian village to find a bank, where we could then get cash by writing a Eurocheck. Eurocheck cards and automated bank teller machines did not widely exist in Europe until later.

The second Bush administration (2001–2008) systematically encouraged the depreciation of the Dollar even though, for the US, the gold standard method of remedying a trade imbalance no longer works. China had pegged the Yuan to the Dollar, guaranteeing that cheap production in China will always win no matter whether the Dollar increases or decreases. The west exports manufacturing eastward, increasing western unemployment and simultaneously increasing inflation via increasing oil prices due to a weak Dollar. At the same time, the US must pay high enough interest rates to attract foreign capital (via sale of US Treasury Bills and bonds) to finance the enormous budget deficit. Via the burgeoning US trade deficit, China had accumulated far more than enough Dollars to offer to buy Unocal for cash in 2005, but the free market US Congress nixed the deal for security reasons. The story of trade with China now is similar to the story with Europe in the 1950s, but for an entirely different reason: the USA sacrifices its currency and loses its productive capacity based on the illusion that free trade produces a social optimum (this is still taught in economics classes).[7] This is the story of the US trade deficit. It's as if Asia has followed the advice of Jane Jacobs to replace imports with its own production, while the west has followed Milton Friedman and the Chicago School of Economics (Friedman, 1975; Friedman and Friedman, 1990) to leave production to those who can produce most cheaply. Friedman ignored the fact that the deregulation/privatization philosophy makes an assumption of *liquidity* of creativity and creation of new industries that fails in reality: inventiveness, new industry, and new jobs cannot arise at a rate fast enough to match the loss of industry to cheaper labor.

---

[7] In the summer of 2007 Harley-Davidson sent representatives to Germany to try to buy a gear-cutting machine. Of the many small companies that had made the machines in the US, none were left.

The history of globalizing capital can therefore be seen systematically as the history of the increase in liquidity and deregulation internationally, and of the loss of manufacturing capacity in the west due to the freedom of powerful corporations to abandon any home region in search of the cheapest labor, the lowest taxes, and new markets. And unfortunately, in 1999 the Glass–Steagall Act was repealed by Congress, and ill-advisedly signed into law by President Clinton, allowing commercial banks to get back into stock market betting. *The assumption that deregulation and privatization lead to optimal societies is essentially the assumption of neo-classical equilibrium with agents replaced by nations.* Only in September, 2008 did widespread fear of the consequences of this 28-year-long program arise in Washington (www.nytimes.com/2008/09/20/business/20politics.html). But let us take our time on the path to the liquidity drought of September, 2008.

9.8 Deficits, the money supply, and inflation

The Dollar weakened dramatically against the Euro and Yen from 2001–2008, while the trade imbalance and budget deficits exploded. A solution is made much harder than in 1982 because the USA has lost too much of its production capacity to Asia, while consuming energy at a very high rate and importing two-thirds of its energy demands (China is similar in this respect). The USA in 2008 consumes more than twice as much energy as either Germany or China.

George Soros asserted in the news in January, 2008 that the status of the Dollar as default worldwide reserve currency has ended. In order to understand better how the USA arrived at a Dollar crisis in 2008, we must review the fiscal/taxation policies under Reagan (1980–1988) and Bush (2001–2008), and the expansion of the Dollar supply shown in Figure 9.1. First, we exhibit some standard academic economic reasoning by reviewing and commenting on the picture of the relation of budget deficits to increases in the money supply provided by Sargent (1986), a rational expectations theorist.

Inflation means an increase in the price level measured by some basket of goods and services. Fiat money, money printed by the Treasury without any backing through private lending, is inflationary since it increases the money supply without increasing the supply of goods and services. The money supply is also increased by credit, which can increase inflation, since via credit more money is funneled into consumption.

To try to quantify the ideas in a rough, elementary way, let $G(t)$ and $Tax(t)$ denote government expenditures and expected tax revenues respectively. If governments operated from taxation alone, then the budget constraint would be $G(t) = Tax(t)$. This was rarely if ever the case. Governments borrowed historically to finance wars and explorations (Columbus's voyages to Ireland, Iceland, and America, for example), so that bankers like the Fuggers in Augsburg, Germany, and later the Rothschilds in London, England, grew in importance to governments as governmental projects became larger and more expensive, and financial markets became correspondingly more powerful. We follow Sargent (1986) in part in what follows.

The so-called "Ricardian regime" is defined as $G(t) - Tax(t) = D(t)$ where debt $D(t) > 0$ is privately financed, there is no new creation of money. For example, with the budget constraint

$$G(t) - \text{Tax}(t) = B(t) - B(t-T)(1 + r(t-T)) \tag{9.4}$$

we can think of government debt as financed by issuing bonds $B(t)$ paying interest at rate r. State and city governments are forced to operate in the Ricardian regime because they cannot "coin" money, and (since the last third of the nineteenth century in the USA) neither can commercial banks. Clearly, the interest rate must be high enough to attract investors to take the risk. But a central government can inflate, can print fiat money $M(t)$, so that

$$G(t) - \text{Tax}(t) = M(t) - M(t-T) + B(t) - B(t-T)(1 + r(t-T)) \tag{9.5}$$

hence financing the debt in part by inflation (generally adiabatically, meaning very slow inflation in order not to disturb the system too much), and in part by private borrowing on the open market via $B(t)$.

Were a government to finance the debt entirely or largely by fiat money, then that government might lose credibility. The German response to the French invasion of the Ruhr coalfields (as demand for payment of war reparations) in 1923 was hyperinflation, $G(t) \approx M(t)$. The money was printed as a form of passive resistance to the occupation and paid out to miners on strike. In this case, Berlin managed to recover its financially reliable image after the French left the Ruhr, and the hyperinflation was ended. Because of Eurodollars, included in M3, Washington does not have the control over the Dollar that Berlin had over the Reichsmark in 1923.

The stated ideological aim of both Reagan (1981–1988) and Bush (2001–2008) economics, was to reduce both $G(t)$ and $Tax(t)$ simultaneously. In both regimes, in contrast to the promised policy, $G(t)$ increased significantly while tax rates were reduced. An implicit liquidity assumption is made in (9.5), which is only a back-of-the-envelope equation that neglects dynamics: one would expect that, with $Tax(t)$ nonincreasing, the government would need to increase interest rates r in $B(t)$ in order to attract investors to finance government debt $G - Tax$. In January, 2008, due to fear of a recession, and on the heels of the subprime mortgage fiasco, the Bush administration proposed reducing both taxes and interest rates simultaneously, with the Dollar already having sunk to a new low of about \$1.50/Euro, and oil correspondingly having hit nearly \$100/barrel. With too few private investors due to too small an interest rate r (compared with the Euro, for example), $M(t)$ would need to be increased via fiat to take up the slack, which would only cause Washington to lose more credibility by decreasing the value of the massive number of Dollars held in Beijing and Moscow, for example. The following argument is a monetarist view (Sargent and Wallace, 1986) of Reaganomics.

In the Reagan and Bush regimes, $G(t)$ increased dramatically while tax rates were systematically and considerably reduced. One possibility was that $M(t)$ could have increased; fiat money could have been printed to pay for the deficit. But this policy would have angered the 10% who disproportionately hold most of the wealth and would have caused foreigners who finance the budget deficit to dump US bonds and Treasury Bills. In an attempt to explain the contradictory policy of increasing $G(t)$ while decreasing $Tax(t)$, Wallace (Sargent, 1986) suggested an analogy with the game of chicken. In chicken, two cars drive toward each other and at the last second one must chicken out, otherwise both drivers die. The one who doesn't chicken out "wins." This is a Nash equilibrium game where self-interest represents the winning strategy, as in all Nash equilibria. Wallace compared the monetary authority (controlling $M(t)$) and fiscal authority (controlling $G(t)$) with the two drivers. Wallace suggests that the ideological aim of Reagan was to force downsizing government by reducing $Tax(t)$, to force the fiscal authority to reduce $G(t)$. This argument is exceedingly naive because it ignores the obvious alternative: to finance the debt privately, borrowing in finance markets, which is exactly what happened. In other words, the chicken game argument was concocted by rational expectations theorists in agreement with the illusion that Reaganomics was supposed to have "forced" the elimination of government programs by eliminating their funding source (taxes). Indeed, the goal of neo-conservatives is to privatize everything, including the schools and (already partly accomplished) the army. In the neo-conservative (neo-con) philosophy, fathered by William Kristol but grandfathered by William F. Buckley, taxation and regulations for "the public good" are oversimplified as socialistic evils that must be eliminated come hell or high water.

Were the budget constraint (9.5) the whole story, there should have been no significant inflation, because the US debt has apparently been largely privately financed, mainly by selling US government debt to Japan and China in the most recent years. Why does the US currently experience an inflation comparable to that of the 1970s? The cost of oil reflects the weakness of the

9.8 Deficits, the money supply, and inflation

Dollar, which reflects agents' lack of trust in Washington's economic and financial policies, including fear over the size of both the budget and trade deficits. On top of that, interest rates have been reduced since 2001 for fear of a recession, making the US debt (Treasury Bills and bonds) less attractive for foreign money, thereby weakening the Dollar further. Oil cost increases also drive up other costs since, for example, agriculture is now centralized (family farms are a "set of measure zero") so that food must be shipped over long distances. With reduced manufacturing capacity, US consumers must be stimulated to spend in order to avoid high unemployment in the service and housing sectors. Hence the housing boom, which was fueled by low interest rates and exceedingly easy credit (but with sliding mortgage rates). Much of the money from which the housing credit was created came from foreign banks, which had encouraged their clients to invest in the US housing boom (this is why the European Central Bank also provided liquidity in 2007 and 2008).

The simple budget constraint (9.5) considers only fiat money created to pay for deficits, money simply printed without taxation or borrowing to cover the amount, and ignores the creation of money via credit. That budget constraint is too simple to describe the inflation of the US Dollar 2001–2008: the money supply in (9.5) represents only an insignificant component of money creation; it does not include M1 and M2, where the credit created is regulated by the Fed, nor does it contain M3.

The component M3 includes so-called "Eurodollars," Dollar deposits and credit created from Dollar deposits outside the US and therefore beyond the control of the Fed. Those Dollars, like M0–M2, are used to create new money via a multiplier effect, but the multiplier is decided by the central bank of the nation where the Eurodollars are deposited. This reflects the role of the Dollar as the international reserve currency. In 2006, the US Federal Reserve announced that it would no longer provide information on M3, but a few private organizations listed on the internet continue that service for the general public (http://seekingalpha.com/article/21027-the-return-of-m3-money-supply-reporting).

Although the Dollar has fallen due to low interest rates and lack of confidence in US governmental policy ("value," p_c, fell by over 60% against the Euro from 2000 to 2008), it seems doubtful that another single currency like the Euro or Yen will replace the Dollar as worldwide reserve currency in the future. The reason is simple: if, for example, the Euro were to replace the Dollar, then the European Central Bank would no longer control that currency entirely, which (like the Dollar) would necessarily grow significantly in quantity (via credit) in the form of M3 in order to finance worldwide economic expansion and consumption. Euros accumulate in Russia for the same reason that Dollars accumulated in OPEC countries in the 1970s: oil and gas purchases in Russia by Europe. In December, 2007 the Euro M3 was 8.7 trillion Euros and, given the exchange rate, roughly matched the Dollar M3 of about 13 trillion Dollars. However, whereas the Euro M2 at 7 trillion Euros accounts for most of the M3, the Dollar M3 is nearly 50% larger than the Dollar M2. So "Euroeuro" reserves would have to increase by several trillion in order to replace the Dollar as international reserve currency at this stage, allowing M3 to become of the same order of magnitude as M2. So far, for the Euro, the M3/M2 ratio has increased in 2007 only slightly above its 110% 2005 ratio. With the second largest Dollar reserves in the world, the Russians started switching from the Dollar to the Euro as reserve currency in 2004. That the Euro M3 has not exploded means that Europe (primarily Germany) has approximately held its own in the export game, while the US has faltered.

Europeans who may take pride in the idea of OPEC, Russia, and China swapping Dollars for Euros have not thought of the possible implications for themselves of an increase in their currency by the creation of trillions of Euros via Euroeuro loans in banks outside of Europe. But, then, most voters do not possess the knowledge necessary to form opinions on the basis of severely idealized notion of "rational expectations" (Chapter 10). Most likely, the role of providing worldwide credit will have to be shared among several currencies in the future, but this would require some sort of agreement to prevent the expansion in M3 created since 1971 for Dollars. That is, agreement on and enforcement of new and more restrictive international trade and monetary regulations will be required in order to reduce both international and local financial instability.

The discussion above is terribly incomplete. We've written as if money creation were due to fractional reserve banking alone, focusing on M2 and M3 which include regulated forms of money creation. We've neglected the most significant contribution of financial deregulation to the world: the role of derivatives in "shadow banking" (Gross, 2007a).

9.9 Derivatives and shadow banking

The Dollar first reached crisis stage c. 1961 when Eurodollars were at least on the order of magnitude of the US gold supply. A new crisis stage is suggested because M3 is roughly twice M2, and the USA has no way under deregulation rules to plug that dike. We can speculate that the next depression could occur when derivatives bets are on the order of magnitude of M2. When that

9.9 Derivatives and shadow banking

occurs, then there will be no way to provide liquidity other than by degrading the currency. Actually, the crisis conditions were already met in 2007. But, we could not predict that the crisis would explode exactly in September, 2008. Recall that crisis conditions were met in 1961, but band aids "worked" until 1971 when the Dollar had to be cut loose from gold. Now, there is no gold to cut loose from, there is only the possibility of money creation either by borrowing, or by fiat, in order to provide liquidity.

The current Dollar credit crisis began in earnest in 2007, and many Dollars and Euros have been created in order to provide the liquidity needed to avoid bank collapses. In the subprime mortgage crash, which surfaced in August, 2007, subprime mortgages were about a trillion Dollars, and total mortgages were about 9 trillion Dollars. How could total mortgages in Dollars be on the order of magnitude of M3? Apparently, excessive amounts of money as credit are being created that do not appear on the balance sheets of any bank. The money is created by derivatives called CMOs. Deloitte (2007) states that worldwide, central banks provided half a trillion Dollars in liquidity in August, 2007, and that was only the beginning of the crisis. In 2008, another 100 billion was provided *by the US Government* to bankrupt AIG, and a consortium of European nations injected another 200 billion in liquidity to prevent further bank collapses due to European bets placed in the American housing market. The subprime mortgage debt is not counted in either M2 or M3 because of the new and powerful role played in international finance by "shadow banking." The mortgages may have been created originally by banks, but were then immediately sold to third-party investors or repackaged into structured products and then sold. By this trick they don't appear on the balance sheet of any bank. The creation of unregulated CMOs and other derivatives faces us squarely with the old problem of the nineteenth century before banks were prohibited from printing their own local currencies. The US Federal Reserve Bank, under Alan Greenspan, systematically adhered to *laissez faire* policies and failed to consider the problem of regulating derivatives markets. Complexity enters because the "structured instruments" created are not understood (Gross, 2007b). Complexity also enters because, while we understand normal liquid FX markets pretty well, we cannot always foresee when a liquidity problem will arise in one market and cause liquidity problems in other markets. But we can always expect that when the bets are on the order of magnitude of the money supply, a crisis is at hand. Clearly, derivatives in particular and finance markets in general should be regulated to prohibit bets of that order of magnitude. The question is: how? With modern technology and quants, hedge funds systematically create ways to escape governmental regulation. International cooperation of governments will be required to prevent brokerage operations from escaping outside national boundaries as a means of avoiding regulating finance and controlling the money supply. Otherwise, as stated earlier, we're in the position of the nineteenth century when commercial banks could print their own paper money. True to the cause of their extremist anti-regulation philosophy, the "Austrian School of Economics" (which has nothing to do with Austria today) is against central bank control of a currency. That is, any instability imaginable is preferred over any reasonable degree of control (but only in theory, of course).

The credit bubble problem is actually far worse than has been described above in the discussion of CMOs. I therefore end with some information from Soros's (2008) book. We've mentioned that M2 is about half M3, which was about $13 trillion in 2007. This includes all "on balance sheet" credit created in Dollars in the world. For reference, US household wealth is about $43 trillion, the capitalization of the US stock markets is about $19 trillion, and the US treasuries market is roughly $5 trillion. Soros (2008) discusses a host of derivatives, all "off balance sheet" including collateralized debt obligations (CDOs), CDO^2s, CDO^3s, and credit default swap (CDS) contracts. The estimated nominal value of CDS contracts outstanding is about $43 trillion. According to Soros (2008) early warnings were given: Greenspan chose to ignore warnings given privately before 2000 about adjustable rate and subprime mortgages, Kindleberger warned in 2002 that there was a housing bubble (no mathematical model was needed or relied on), and Volcker and others voiced warnings. Soros (2008) states that hedge funds who tried to sell housing short before the bubble popped suffered bad enough margin calls that they quit betting against the bubble, but after the AIG/Lehman Brothers liquidity crisis of September, 2008, a continuation of the subprime mortgage fiasco, shorting bank stocks was temporarily banned (the first sign of finance market regulation to come).

Standard economists, bureaucrats, and politicians have sold the idea of deregulation as if it would achieve stability. There was no ground whatsoever for that belief other than the mathematical delusions created by too-respected economics theorists like Lucas and his predecessors, based on the empirically untenable assumptions of equilibrium and optimizing behavior. Osborne (1977) knew better. Why did rational expectations grow so in influence? Because it coincided perfectly with the dominant illusion (born 1981, died 2008) that deregulation would solve all economic problems.

Bubbles have been discussed in the literature (MacKay (1980) and Kindleberger (1996)) but since 1973 there's a new element that was not present earlier: the reliance on mathematical models in trading strategies. The basic theory of a single option based on a stock is simple. But some derivatives are options on other complicated options, and mathematizing that is nontrivial. Predictions based on models of derivatives linked together are not reliable. The empirical data needed to falsify such models is simply not available.

9.10 Theory of value under instability

Hume's (1752) price-specie flow mechanism described a tendency toward equilibrium in the marketplace and in FX transactions on the gold standard. If a country imported too much, then the currency piled up in the hands of foreigners, reducing the money in circulation inside the economically weaker country. Prices fell, manufacturing picked up as foreigners bought exports, and the local currency began to increase inside the financially weaker country as prices rose. In that era, factories were not exported outside the US, workers were simply laid off temporarily and then rehired as production picked up again (as in the coal mines and auto factories of the 1950s and 1960s). This was a regular feature of American factory life until c. 1970.

In the absence of a gold standard (which ties money to a particular physical unit, namely, mass), what is a currency worth? What determines its "value"? The theory of value in nonstationary markets is as follows: a currency has no inherent value, its value is whatever price FX traders believe the currency to be worth. What determines the traders' beliefs? If a country produces high-quality and attractive products cheaply enough, then exports will increase. Foreigners will want that country's currency in order to buy the exports. If, on the other hand, a country imports more than it exports then the currency piles up (as Eurodollars, for example) outside the country. But exports do not become cheaper, and the currency correspondingly does not rebound by this scheme because too much production has been moved outside the USA. Hume's equilibrium theory of value no longer applies, the Dollar simply falls persistently in value against stronger currencies as imports are relied on irreplaceably. Exports can't stem the flood of Dollars outside the USA unless the country produces and sells enough internationally to match the flood of imports. The factories that would have produced the exports no longer exist inside the USA, and cannot be rebuilt without enormous time and investment (this is a like a liquidity drought). Budget deficits magnify the instability of a currency under these circumstances: when a government spends too much and taxes too little, then the money may be borrowed from the same foreigners who have accumulated the currency due to the export-import imbalance. This weakens the currency further in the minds of speculators, who easily sense the instability.

Like 1932 and 1981, the financial crash of 2007–2008 is a turning point in ideas about political economy. After seeing the free market idea explode into absurdity in September, 2008, many deregulation enthusiasts finally began to see that markets require regulation if we're to avoid global liquidity crises. In Germany, 480 billion Euros was guaranteed by the government to prevent bank collapses. In the US, more than a trillion dollars has been thrown at the financial system, with partial government ownership of banks as a consequence. It's exceedingly shortsighted, as former Federal Reserve Chairman Greenspan does, to blame the financial collapse on greed and to continue to argue that derivatives trading should not be restricted. Here's my last analogy. Deregulators are typically strong believers in "law and order." In the case of street crime and ordinary burglary, they believe that strong laws with stiff penalties can effectively deter armed robbery and other street crime. It's inconsistent to fail to apply the same idea to finance markets and trading, to believe that the humans in Wall Street markets are more moral than the humans on the Main Street markets and in the back alleys. Stated otherwise, mathematical theorizing about market preferences/utilities alone idealizes human behavior in a way that ignores avarice, and therefore cannot be helpful.

9.11 How may regulations change the market?

In September, 2008 the USA began again to regulate financial markets. Short selling of financial stocks was (at least temporarily) outlawed, and the failed mortgage betting debt of finance houses may be bought by the US government. The total of all bailouts is expected to amount to more than a trillion Dollars. We can compare this with the annual budget deficit of half a trillion with interest payments. How finance markets will be affected by new regulations depends on how the bailouts are financed, and whether regulations effective enough to prevent bets on the order of magnitude of the money supply are constructed. So far as financing bailouts, there are three alternatives, and combinations thereof: (i) to raise taxes, (ii) to borrow in international finance markets, or (iii) to print fiat money. Unless taxes are raised and trade tariffs are imposed to rebuild industry, we should expect a much greater inflation than has been experienced to date.

Market regulations may well reduce financial trading, but statistical equilibrium will not emerge. If leveraged derivatives trading were outlawed then liquidity droughts should become much less likely. We won't know for at least five to six years after the institution of regulations how the market dynamics will be changed (see Chapter 7 for the limitations on discovering dynamics from sparse data). We can expect that volatile Martingale dynamics will persist, that markets will remain nonstationary/unstable (because there can be no return to anything equivalent to a gold standard). Volatility would be reduced significantly by a return to something like the Bretton Woods Agreement, but that is made difficult by the inherent instability of money and governmental economic policies.

The traditional monetarist/rational expectations viewpoint on money supply growth is discussed in the next section. Rational expectations theorizing was in its heyday in the 1980s and likely to be made defunct by the 2007–08 financial crisis, if not by the analysis presented in this book. Predicting realistically and usefully how fast the money supply should grow is still unsolved, and the entire discussion is made useless so long as money creation via derivatives is left unregulated.

10

Macroeconomics and econometrics

Regression models vs empirically based modeling

10.1 Introduction

The level of business activity depends strongly on the ease of obtaining money as credit, and the level of money should in principle be determined by central banks.[1] However, central banks rely notoriously on models, and the economic models used have not been empirically deduced from observed time series. Rather, the models have been postulated, or "validated," on the basis of regression analysis, while relying on an assumption of stationarity of economic variables. In later, more sophisticated treatments, the method of cointegration, which assumes stationary increments and also ergodicity, is used. The assumptions made by economists about the noise in the regression models cannot be justified empirically, and the models yield spurious predictions of stability.

Keynes stated that economic theory is used for creating and maintaining economic policy and models have been invented that support one brand of economic policy or another. Keynesian economics encouraged government intervention to try to fine-tune the economy from the Great Depression until at least the early 1970s. The emergence of rational expectations in the 1960s and beyond provided theoretical support for *laissez faire* policies (Bell and Kristol, 1981). Deregulation as policy has dominated in the USA and UK since at least 1981, and has spread worldwide on the advice of economists based on rational expectations theory. The aim here is to expose the holes in the rational expectations claims, and to see that that model does not provide any empirical basis for adopting *laissez faire* as policy. Rational expectations exemplifies what postmodernists would label as "socially constructed theory."

---

[1] This viewpoint ignores derivatives and off-balance-sheet money creation.

10.1 Introduction

Keynesian ideas of governmental intervention became popular during the liquidity drought of the 1930s when deflation and high unemployment were the main economic problems. Fiscal policies were instituted both in Germany and the USA to employ jobless people. Money was in short supply, and those who had money didn't spend enough to alleviate the high unemployment. The Keynesian advice made sense: when money is in too short supply, consumption can be stimulated either via taxation and redistribution, or by printing or borrowing money and redistributing it to those who need it (neo-classical economics ignores "needs" in favor of "preferences"). Keynesian macroeconomics was geared to specific historic conditions and was unable to deal successfully with the 1970s stagflation (inflation combined with unemployment). We understand why inflation in that era was high: because of the large number of Eurodollars relative to gold, the Dollar was deregulated from the gold standard in 1971. Runaway inflation inside the USA was the result. In that era, monetarist policy based on rational expectations theory emerged and dominated until October, 2008.

Rational expectations was created in an era when shadow banking and the flood of derivatives did not exist. Rational expectations can properly treat neither money nor derivatives, but the underlying economic philosophy of "hands-off business" promoted the unregulated explosion of credit via derivatives that led to the bust of 2007–08. A mathematical presentation of rational expectations theory, from its neo-classical foundations through regression models, can be found in Sargent (1987).

In order to go to the source of the confusion reigning in economic theory, we begin with Muth's original model of rational expectations (1961, 1982). We will expose the inconsistency in Muth's derivation. The rational expectations model's policy predictions are discussed in Section 10.5. In Section 10.6 a more realistic model of macroeconomic behavior is presented. There, we replace the untenable assumption of stationary economic variables by non-stationary ones. Throughout, we will use models continuous in price and time for mathematical convenience. Our viewpoint could be reformulated to describe more realistic discrete models, but too much work would be required for too little payoff. Now and then, however, we will pay attention to the fact that prices are stated at most to three decimal places.

The reader should understand that regression analysis in statistics originally was a method whereby theoretical predictions could be compared with empirical measurements. The theory was either assumed to be correct (e.g. describing a planetary orbit), or else one wanted to test a calculation, and the difference between theory and measurement was attributed to "scatter" due to finite precision. Although the scatter is sometimes called "error," this is not theoretically correct: Tschebychev's Theorem predicts that scatter *must* be present because we cannot carry out infinitely many reruns of an experiment of observation. The so-called "error" was therefore assumed, with good reason, to be stationary. Here's our first point: in econometrics, the regression model *becomes* the theory. *The assumption that the noise in that case is stationary is untenable.* We will first follow the economists' standard arguments, and then point out exactly where and why stationarity assumptions must be abandoned if we're to present a realistic picture of macroeconomic variables. This path will lead us all the way through cointegration and ARCH/GARCH models to the frontiers of modern econometrics.

10.2 Muth's rational expectations

We rely on the history of neo-classical economic theory in order to understand the unnecessary and wrong steps in Muth's original argument. Muth began (i) by deriving time-dependent prices from an "equilibrium" assumption, and then (ii) replaced subjective expectations of a representative agent whose ensemble average expectations agree with those calculated for the stochastic process under consideration. One is supposed to take point (i) for granted, and then appreciate point (ii). Here's the necessary historic background.

Arrow and Debreu, whose ideas of implicitly stable equilibrium dominated economic theory in Muth's time and beyond (Geanakoplos, 1987), had created a theory where uncertainty/probability was thought to have been banished (McCann, 1994). That banishment was based on the noncomputable and humanly impossible requirement of perfect knowledge and infinite foresight on the part of all agents (see Chapter 2 in this book). The uncertain reality of market statistics was disallowed in favor of absolute certainty in the neo-classical model. Probability was reduced to *subjective* choices among fixed and well-defined alternatives/preferences. It was necessary to pay lip service to those ideas in order to be published. Even today, one generally cannot publish criticism of neo-classical ideas in mainstream economics journals. This is the background needed to understand Muth's artificial-sounding argument (point (ii) above) to replace subjective probabilities by expectations based on the theory of stochastic processes, where he "derives" time-varying prices from an invalid market-clearing assumption (point (i) above).

In the rational expectations model, broad deviation from the Arrow–Debreu program (Geanakoplos, 1987) is avoided by limiting market uncertainty to stationary processes. Stationary processes are absolutely necessary if

10.2 Muth's rational expectations

there's to be any hope at all of maintaining agreement with neo-classical predictions of relations between variables on the average in regression analysis.

Muth's original model of rational expectations is based on a mathematical self-inconsistency committed originally by Ezekiel, but, as strange as it seems, still defended today by neo-classical economists. Consider a market with one item. Assume that there exists an equilibrium price $\bar{p}$. Let $p(t) = \bar{p} + \delta p$. In a deterministic market the equilibrium price is obtained by solving $D(\bar{p}) = S(\bar{p})$ where $D(\bar{p})$ is demand and $S(\bar{p})$ is supply. Following Muth, assume instead that demand/consumption and supply are given by

$$D(\delta p, t) = \beta \delta p(t)$$
$$S(\delta p, t) = \gamma \langle \delta p \rangle_{subj} + u(t) \tag{10.1}$$

where $\langle \delta p \rangle_{subj}$ is supposed to be agents' subjective estimate of future price change made on the basis of past knowledge, and $u(t)$ is random noise to be specified. The noise represents the uncertainty in the market and is the source of liquidity. Setting demand equal to supply yields

$$\delta p(t) = \frac{\gamma}{\beta} \langle \delta p \rangle_{subj} + \frac{1}{\beta} u(t) \tag{10.2}$$

From an assumption of equilibrium is derived a non-time-translationally invariant price. This is a contradiction, but before correcting the mistake let's continue with Muth's so-called derivation of rational expectations.

The process $u(t)$ is assumed stationary, so the price process $p(t)$ is also stationary with time-invariant one-point density $f_1(p)$. Fluctuations about equilibrium are described by a time-translationally invariant two-point density $f_2(p, t + T; p_0, t) = f_2(p, T; p_0, 0)$ and by higher-order translationally invariant densities. Averaging (10.2) using the equilibrium density $f_1(p)$ generated by time series obeying (10.2) yields the ensemble average prediction

$$\langle \delta p(t) \rangle = \frac{\gamma}{\beta} \langle \delta p \rangle_{subj} \tag{10.3}$$

"Rational expectations" amounts to assuming that agents' subjective and ensemble averages agree,

$$\langle \delta p(t) \rangle = \langle \delta p \rangle_{subj} \tag{10.4}$$

In a stationary model this means one of two things. Either (i) $\gamma = \beta$, or (ii) $\langle \delta p(t) \rangle = 0$ so that $\langle p(t) \rangle = \bar{p}$. Since $\delta p = p(t) - \bar{p}$, in either case the prediction of Muth's rational expectations is simply that

$$\langle p(t) \rangle = \bar{p} \tag{10.5}$$

the expected price is the equilibrium price. With an ergodic stationary process, rational expectations claims that the "rationally-subjectively expected price" defined as ensemble average is the same as the equilibrium price computed from the price history of the model. This means that past history can be used to determine future expectations. That the model, in the end, represents a stationary hypothetical market is correct. But the claim that market clearing holds for that hypothetical stationary model is wrong because point (i) above is inconsistent with the statistical ensemble predictions of the stationary market model. We know from Chapter 4 that hypothetical stationary markets would clear only on the average, and that the equilibrium price is agreed on only by a small fraction of all agents. Stated otherwise, hypothetical stationary markets do not clear due to fluctuations.

If we relax the stationarity assumption, as real markets demand, then we can see that "rational expectations" is neither a theory nor a model. Rational expectations should simply mean that the ensemble averages of a chosen model define our expectations about the future, if we extrapolate and assume that the future will resemble the past statistically. From a physicist's perspective, the notion of subjective probabilities was unnecessary from the start. Subjective probabilities may be of interest in agent-based modeling, where the modeling should be constrained by macroeconomic facts, if one can get a clear picture of exactly what are macroeconomic facts from the standpoint of statistical ensembles.

From a practical standpoint, standard rational expectations assumes that the "representative agent" should form expectations based on his or her best guess about the future, updating his or her knowledge up to the present. *But there can be no surprisingly new knowledge in a stationary market, because recurrence ensures us that the future is statistically a repetition of the past* (Kac (1949) showed that all discrete stationary processes are recurrent, even with no assumption of ergodicity). The words about agents updating their knowledge in order better to anticipate the future actually acknowledge the fact that real markets are nonstationary, but those words are relatively empty in the context of hypothetical stationary models of economic variables. The only "updating" that can be done is to place a buy order if prices fall far enough below value, and a sell order when the reverse occurs, profitable fluctuations being guaranteed by recurrence. Presumably, by using "the best forecast," the rational expectations advocates likely meant that the representative agent should optimize one or another ad hoc expected utility. But traders and financial engineers do not waste time on utility-based models when using or creating synthetic options.

The advice to use the best possible forecast in a real, nonstationary world is imprecise and ill-defined. What does a forecast mean in that case? How should agents know how to distinguish knowledge from noise, and to arrive at "the best forecast of the future"? A "best forecast of the future" is not necessarily the same as using a historic time average to predict the future. The latter does not take new knowledge into account; it extrapolates by using the assumption that the future resembles the past, which is what's generally assumed by insurance companies and financial engineers for real, nonstationary processes.

As Arrow and Debreu may have anticipated in their adherence to subjective probabilities on sets of fixed alternatives, the neo-classical picture begins to dissolve when faced with uncertainty. The assumptions of maximizing one or another expected utility on the one hand, and statistical equilibrium on the other, may be retained by ivory tower theorists, but all features of that worldview disintegrate in the face of market reality. When Muth assumed that supply and demand match in real time by using (10.1) to obtain (10.2), he created a mathematical contradiction. When supply and demand match for all times then markets clear, and 100% of all traders agree on exactly the same price. No trading takes place in the neo-classical world until the equilibrium price is established by the Walras auctioneer, or by the central authority, neither of which is computable. *In an uncertain but still stationary world, markets do not and cannot clear.* Even hypothetical stationary markets cannot clear. In the neo-classical mindset, trading should not take place at all in a rational expectations model because even stationary fluctuating prices are out of equilibrium.

10.3 Rational expectations in stationary markets

The assumed basis for Muth's model, market clearing with $D = S$, is wrong because time-translational invariance (equilibrium) does not tolerate a time-varying price. We can formulate the model correctly, first for hypothetical stationary markets, and later for nonstationary ones. Demand vs supply means that

$$\frac{dp}{dt} = D(p,t) - S(p,t) \tag{10.6}$$

so that market clearing is possible if and only if the price is constant and solves $D(p) = S(p)$. This determines the equilibrium price $\bar{p}$. In a stochastic model, market clearing cannot occur because of uncertainty. A stochastic model with an approach to equilibrium is given by the OU model

$$dp = -rp\,dt + \sigma_1 dB \tag{10.7}$$

with $r > 0$ where $B(t)$ is the Wiener process and $u(t) = dB/dt$ is white noise. Here, the right-hand side of (10.7) represents excess demand $D - S$ in an uncertain market. At best, in statistical equilibrium (at long enough times $rt \gg 1$) we can obtain vanishing of excess demand on the average,

$$\left\langle \frac{dp}{dt} \right\rangle = 0 \qquad (10.8)$$

That is, in an uncertain market, the market can never clear in detail but clears on the average. One way to remedy the Ezekiel–Muth error of deriving a time-varying price from market clearing is, instead of (10.2), to write

$$\frac{dp}{dt} = D(p,t) - S(p,t) = \beta p(t) - \gamma \langle p(t) \rangle + u(t) \qquad (10.9)$$

where rational expectations requires $\gamma = \beta$. Statistical equilibrium exists if and only if $\beta = \gamma < 0$, reflecting negative demand. If we interpret "$\langle p(t) \rangle$" in (10.9) not as the average price at time t but rather as the equilibrium price (the average price when $\beta t \gg 1$), then the model is, to within a shift of variable, the OU model (10.7). But negative demand is not of economic interest. Therefore we must find a better way to derive Muth's stationary model.

Here's the mathematically correct solution. Assume a market in statistical equilibrium, and assume we don't care how it got there. That is, we do not prescribe a model like (10.9) of how equilibrium is reached from a non-equilibrium state (this is analogous to ignoring the role of the Walras auctioneer or central authority in neo-classical economics). Equilibrium values are constant averages $\langle p \rangle$, $\sigma_\infty^2 = \langle p^2 \rangle - \langle p \rangle^2$, etc., calculated from a stationary price density $f_1(p)$, $\langle p^n \rangle = \int_0^\infty dp\, p^n f_1(p)$, and the corresponding stationary stochastic process

$$p(t) = \langle p \rangle + u(t) \qquad (10.10)$$

with $\langle p \rangle = \bar{p} = \int dp\, p f_1(p)$, describes fluctuations about statistical equilibrium. A simple example of uncorrelated stationary noise is what economists and statisticians label "white noise," a Gaussian process where $\langle u(t)u(s) \rangle = 0$ if $s \neq t$) with constant variance $\langle u^2(t) \rangle = \sigma_1^2$, but "$u(t)$" is identified as an increment of a process.

The average price can be identified as time-invariant "value." In contrast with the neo-classical barter model where 100% of all agents agree on "value" and have infinite foresight, only a small fraction of our hypothetical agents agree on "value" because of market uncertainty. If the density $f_1(p)$ is approximately symmetric, then the average and most probable prices coincide and we can take the most probable price p_c, the consensus price, as "value."

10.3 Rational expectations in stationary markets

The most probable price p_c locates the peak of $f_1(p)$. In a stationary market, only the fraction of agents described by the small region near the peak of the distribution agree on "value," but overvalued and undervalued are defined in a precise, time-invariant way. Statistical certainty about the future is restored (for discrete prices) by recurrence, so that derivatives and hedging are unnecessary.

Stationary regression models with lag times can now be considered. Assuming a stationary process $u(t)$, assume, for example, that

$$\langle p(t) \rangle_{\text{cond}} = \lambda p(t-T) \tag{10.11}$$

where we take the expected price at time t to be proportional to the last observed price $p(t-T)$ one period earlier. The assumption (10.11) yields the stochastic difference equation

$$p(t) = \lambda p(t-T) + u(t) \tag{10.12}$$

where the conditional expectation (10.11) is computed using the transition density $p_2(p,T|p_0,0)$ for the stationary price process. With noise and price assumed uncorrelated, $\langle u(t+T)u(t) \rangle = 0$, $\langle p(t)u(t) \rangle = 0$, we obtain the unconditioned average

$$\langle p(t)p(t-T) \rangle = \lambda \langle p^2(t-T) \rangle \tag{10.13}$$

For a *stationary* process the pair correlations can depend only on time lag T, not on the observation time t, so that $p(t)$ is stationary only if the variance is constant, $\langle p^2(t) \rangle = $ constant, yielding pair correlations (10.13) that also are constant $\neq 0$ and thus violate the EMH. With $\langle u^2(t) \rangle = \sigma_1^2$ constant, the constant variance $\sigma^2 = \langle p^2(t) \rangle - \langle p(t) \rangle^2$ satisfies

$$\sigma^2 = \frac{\sigma_1^2}{1-\lambda^2}. \tag{10.14}$$

We obtain a stationary model for $0 \leq \lambda < 1$ with a singularity at $\lambda = 1$. That singularity arises from the fact that a Martingale condition

$$\langle p(t) \rangle_{\text{cond}} = p(t-T), \tag{10.15}$$

the assumption of an efficient market, cannot be represented by a stationary market. Economists who believe that stationary markets are efficient simply have not considered the pair correlations.

In standard rational expectations modeling it's largely irrelevant which particular distribution is used to model the noise; the important assumption above is that noise and prices are *stationary*. Rational expectations, from our standpoint, does not require optimizing behavior on the part of the agents, although the two principles of equilibrium and optimization are regarded as reflecting the standard ideology (Lucas, 1972). Optimizing behavior does not produce market efficiency; market efficiency (the Martingale condition) implies market instability.

10.4 Toy models of monetary policy

We reiterate our viewpoint on regression analysis, or curve fitting. Physicists also use regression analysis. For example, if we know or expect that the data are described by a specific theory, then we may make a regression analysis in order to try to fit empirical data. Compared with theory, data will always show scatter due to finite precision even if there would be no measurement error (e.g., due to a finite-sized probe in a turbulent flow). For example, if we know that the measured variable y should be a parabola, then we would write $y = a + bx + cx^2 + \varepsilon$ where ε is assumed to describe the scatter and can be taken to be stationary. In economics, we have no idea *a priori* what the model should be; we have to discover it from the data. In econometrics, a regression analysis with stationary "error" (market noise) *becomes* the model. This is quite different than starting with a known, correct theory as in physics, or with a model deduced empirically from the data as we've exhibited in Chapter 7, where a Martingale was established via lack of increment correlations, so that the theoretical prediction (7.33) was established. In other words, regression analysis and theory are confused together in economics/econometrics, and we must separate one from the other in order to get any idea what's going on. The most modern and most advanced development in regression analysis in econometrics is called "cointegration," which is also described below.

We've pointed out in the last section that there is triviality in discussing "the best forecast" in a hypothetical stationary market. We must ignore that fact for now in order to try to understand the basis for Lucas's policy neutrality advice. Here, we will follow the readable paper by Sargent and Wallace (1976), and the book by McCallum (1989).

Let $m(t)$ denote, for example, the logarithm of the money supply and let $y(t)$ denote a variable that one wants to forecast or control. If $y(t)$ is an interest rate, a return, then $y(t)$ is also the log of a "price." We first assume that all processes under consideration are stationary. Keynesians were motivated to invent econometric models in order to try to defend their policy advice from a scientific standpoint. The monetarists rightfully criticized those models, which did not work, and went on to replace them by models reflecting their own ideology.

In the Keynesian era, the *laissez faire* monetarist Milton Friedman proposed the notion of increasing the money supply at a fixed rate of x% per

10.4 Toy models of monetary policy

year. Keynesians could argue as follows that Friedman's rule is not optimal for setting policy. Let $y(t)$ denote a variable that the central bank wants to control, and assume that

$$y(t) = \alpha + \lambda y(t-T) + \beta m(t) + u(t) \tag{10.16}$$

That is, rational expectations assumes simply that processes are linearly related (as in standard regression analysis) and that all processes considered are stationary. In particular, $u(t)$ is assumed stationary with no pair correlations, and is assumed uncorrelated with $y(t)$. With the idea in mind of (i) setting average $\langle y(t) \rangle$ equal to a target y^*, and (ii) minimizing the variance about that target, assume a linear feedback rule

$$m(t) = g_0 + g_1 y(t-T) \tag{10.17}$$

This yields

$$y(t) = A + \lambda' y(t-T) + u(t) \tag{10.18}$$

where $A = \alpha + \beta g_0$ and $\lambda' = \lambda + \beta g_1$. We then obtain

$$\langle y(t) \rangle = \frac{A}{1 - \lambda'} \tag{10.19}$$

and

$$\langle y^2(t) \rangle = \frac{A^2 + \sigma_1^2}{1 - (\lambda')^2} \tag{10.20}$$

Setting the expected value equal to the target value, $\langle y(t) \rangle = y^*$ fixes g_0,

$$y(t) = y^* + u(t) \tag{10.21}$$

This guarantees that the fluctuations $\delta y = y(t) - y^*$ are uncorrelated, $\langle \delta y(t+T) \delta y(t) \rangle = 0$.

We can minimize the variance $\sigma^2 = \langle y^2(t) \rangle - \langle y(t) \rangle^2$,

$$\sigma^2 = \frac{\sigma_1^2}{1 - (\lambda')^2} \tag{10.22}$$

by taking $\lambda' = 0$, or $g_1 = -\lambda/\beta$. Since Friedman's feedback rule is given by $g_1 = 0$, that rule is suboptimal in the sense that the variance about the target is not minimized. This provides an example of what is meant by producing an optimal forecast in a stationary world. But there's a serious contradiction with reality that we must eventually face: in a stationary world nothing can grow, including the money supply. Before generalizing to nonstationary variables we exhibit the class of models behind Lucas's *laissez faire* policy advice on the basis of rational expectations's stationary world.

10.5 The monetarist argument against government intervention

Lucas's aim was apparently twofold: (i) to poke holes in Keynesian policy analysis, and (ii) to see in his critique the generalization to deduce *laissez faire* as the logical consequence, that government intervention is by mathematical necessity ineffective. The first aim is admirable and useful; a scientist should always try to find the weakness in every theory. Indeed, Lucas succeeded there. His second goal was to derive *laissez faire* policy advice as if it would be a mathematical necessity. The reason why the academic economics profession honors Lucas is that, in contrast with Muth who reasoned minimally, Lucas brought in the full, cumbersome apparatus of neo-classical optimizing to "explain" the rational expectations model.

Suppose that instead of (10.16) and (10.17) we would be faced with the processes

$$y(t) = \varsigma_0 + \varsigma_1(m(t) - \langle m(t) \rangle_{\text{cond}}) + \varsigma_2 y(t-T) + u(t) \quad (10.23)$$

and

$$m(t) = g_0 + g_1 y(t-T) + \varepsilon(t) \quad (10.24)$$

so that instead of (10.17) we have

$$\langle m(t) \rangle_{\text{cond}} = g_0 + g_1 y(t-T) \quad (10.25)$$

Substituting (10.24) and (10.25) into (10.23) yields

$$y(t) = \varsigma_0 + \varsigma_2 y(t-T) + \varsigma_1 \varepsilon(t) + u(t) \quad (10.26)$$

which is independent of the parameters g in (10.24), so that an optimal policy for y is independent of g_0 and g_1. This is neither a deep result nor even a surprise; the g were systematically eliminated from $y(t)$ by direct construction of (10.23) and (10.24)! To see more thoroughly the shaky ground on which Lucas's policy neutrality advice is based, the reader is invited to follow the Sargent–Wallace extension of this argument to unemployment. Clearly, from a scientific standpoint there is no ground here for advising *laissez faire*, Keynesian, or any other policy on the basis of existing econometric models because the models are not derived from macroeconomic data.

Although the above equations have no empirical basis whatsoever, Lucas's argument was accepted as indicating that one cannot expect to tame or otherwise influence business cycles (production drops and unemployment) by managing the money supply or other macroeconomic variables. From a scientific standpoint, Lucas's conclusion should have been entirely different: we should investigate the fluctuations of the money supply $m(t)$ and all variables $y(t)$ of interest empirically to find out how they behave statistically, if possible.

The Sargent–Wallace paper repeats Lucas's advice that the government should not try to "lean against the wind." This is nothing other than *laissez faire* advice against government intervention in the economy. It supported the vast deregulation from the Reagan–Thatcher era until the bursting in 2007–08 of the worldwide derivatives-based credit bubble. *Laissez faire* advice was based on stationary models that fail miserably to describe the behavior of real macroeconomic variables. There is no reason to expect that regulations are *a priori* bad or wrong; indeed we see now, as in the 1930s, that regulations are needed in order to reduce the chance of liquidity droughts and depression.

In spite of the stationary models and their vast influence, in theoretical economics it's long been known (but only quietly mentioned in polite company) that macroeconomic variables are nonstationary. The final escape hatch we need to close is called "cointegration." Cointegration recognizes the nonstationarity of macroeconomic variables, but constructs the illusion that stable relationships between unstable variables can be discovered and predicted. We write "illusion," because we will show as an example below that the conditions under which cointegration works are severely violated by finance markets.

10.6 Rational expectations in a nonstationary world

Consider next nonstationary macroeconomic variables. As hypothetical examples, equations (10.16) and (10.17) would then become

$$y(t) = \alpha + \lambda y(t-T) + \beta m(t) + b \bullet \Delta B \tag{10.27}$$

and

$$m(t) = g_0 + g_1 y(t-T) \tag{10.28}$$

where $b \bullet \Delta B$ is the Ito product and $b(y,m,t)$ defines the nature of the noise and is to be empirically determined. Since the variances are not constant, one cannot minimize as before, but a Keynesian can still minimize the volatility: with

$$y(t,T) = \alpha + (\lambda + \beta g_1 - 1)y(t-T) + b \bullet \Delta B \tag{10.29}$$

we have

$$\langle y^2(t,T) \rangle = \alpha^2 + (\lambda + \beta g_1 - 1)^2 \langle y^2(t-T) \rangle + \int_{t-T}^{t} \langle b \rangle^2 dt \tag{10.30}$$

Volatility is minimized if we choose $g_1 = (1-\lambda)/\beta$.

If instead we make (10.23) and (10.24) nonstationary then, as before, we cannot minimize the volatility of $y(t)$ by playing with g_1 unless the diffusion coefficient for $y(t)$ depends on $m(t)$, which is possible. The correct implication is not that business cycles cannot be influenced by monetary policy, but rather that the first order of business would be to try to establish correct empirically based equations of motion for variables $y(t)$ of interest and for the money supply $m(t)$. To date, no correct analysis of money supply fluctuations has been performed. Without a correct empirically based set of equations, macroeconomic theory cannot produce insight into macroeconomic phenomena.

We've seen in Chapter 9 how an equilibrium model can be imposed via strong regulations (stationarity under the gold standard via boundary conditions). That is, a model of social behavior can be enforced. This was also the case under communism, even if we don't like the result. One could argue that, for many years, life under communism was relatively stable and restrictive under dictatorial regulations.

We've shown in Chapter 7 the basic limitations on discovering the dynamics of deregulated FX markets: even a six-year time series is "too short" from the standpoint of the required statistical ensemble. The earlier alternative was to make questionable assumptions, and try to turn regression analysis into a mathematical model. That's exactly what's done under rational expectations, and while econometricians have been regressing, cointegrating, and assuring us that macroeconomics is like physics (we need only calculate and predict), worldwide financial instability has exploded in our faces. *Guessing* how macroeconomic variables might *hypothetically* behave under strong assumptions (regression analysis) is not a substitute for *discovering* how those variables *really* behave (statistical ensemble analysis). And if the data are too sparse, then we cannot, with any degree of statistical probability, claim to understand how the variables behave. We end this section with a telling quote from the 2004 Sveriges Riksbank Lecture:

Macroeconomics has progressed beyond the stage of searching for a theory to the stage of deriving the implications of theory. In this way, macroeconomics has become like the natural sciences. Unlike the natural sciences, though, macroeconomics involves people making decisions based upon what they think will happen, and what will happen depends upon what decisions they make. This means that the concept of equilibrium must be dynamic, and – as we shall see – this dynamism is at the core of modern macroeconomics.

Edward C. Prescott

(http://nobelprize.org/nobel_prizes/economics/laureates/2004/prescott-lecture.html)

10.7 Integration I(d) and cointegration

10.7.1 Definition of integration I(d)

The standard expectation in econometrics and macroeconomic data analysis is that $x(t,-T)$ may be a stationary increment, with fixed T, even if the

10.7 Integration I(d) and cointegration

stochastic process $x(t)$ is nonstationary. Stated explicitly, if the process x is nonstationary then form the first difference $x(t,-T) = x(t) - x(t-T)$. If the first difference $x(t,-T)$ is nonstationary with T fixed, then one may study the second difference $x(t-T, -T) = x(t) - 2x(t-T) + x(t-2T)$, and so on until either stationarity is found or the chase is abandoned. If the process $x(t)$ is already stationary then it's called I(0). If the process $x(t)$ is nonstationary but the first difference $x(t,-T)$ is stationary for fixed T then the process is called integrated of order 1, or I(1). If neither x nor the first difference is stationary but the second difference is stationary for fixed T, then the process is called I(2). According to S. Johansen (1991, 2008) the above prescription is not strictly correct: it's possible to construct a two-variable regression model based on special assumptions about the noise where both the processes x, y and differences $x(t,-1)$, $y(t,-1)$ are nonstationary ($T = 1$ here) but there is still a stationary linear combination $\alpha x(t) + \beta y(t)$.

By "noise" we mean any drift-free stochastic process. Typical examples of noise are the Wiener process, white noise, statistically independent nonstationary noise, iid noise, drift-free stationary processes, Martingales, fBm, and the correlated noise of near-equilibrium statistical physics. "Noise" therefore implies nothing whatsoever about correlations, only that the stochastic process is drift-free.

Our goal in this chapter is to pin down the class of noise processes $x(t)$ for which integration I(1) is possible when T is held fixed. In theoretical discussions of integration and cointegration, either "white noise" or iid noise is assumed ad hoc. In the context of the Granger Representation Theorem, it's been stated that the practitioners of cointegration generally do not worry much about the noise distribution because the cointegration technique is presented primarily as matrix algebra (Hansen, 2005). In cointegration studies the test of the noise distribution generally does not go beyond checking for a (presumably one-point) Gaussian distribution. It's also known that a change of time variable is sometimes adequate to transform nonstationary differences to stationary ones but is generally inadequate, and we will explain below why such a time transformation generally cannot be found.

The tradition in macroeconomics is to postulate the noise in as simple a way as possible ("iid" or "white") instead of discovering the noise distribution from time series analysis. We will therefore analyze the distributions of "white" and iid noise processes below, and will show that stationary increment Martingales include the economists' "white noise." First, we summarize standard economics viewpoints about regression analysis and cointegration in the next two sections.

10.7.2 Regression analysis as a search for stationarity

To illustrate regression models, consider any macroeconomic variable $x(t)$ like unemployment, the price level, the money supply, or an exchange rate. Suppose that by ignoring uncertainty, macroeconomic theory predicts or speculates that $x(t) = \lambda x(t-1)$ should hold, where $t-1$ is the present time and t is one period later. Then the hope is to find that

$$x(t) = \lambda x(t-1) + \varepsilon(t) \tag{10.31}$$

where $\varepsilon(t)$ is drift-free uncorrelated noise with zero mean and finite variance (modulo fat tails), and λ is a free parameter. This defines what econometricians mean by "white noise" if the variance is taken to be constant. *If the noise is stationary then so is $x(t)$ if there is no drift in (10.31).* Assuming "white noise" $\varepsilon(t)$, we obtain

$$\langle x(t) \rangle = 0$$
$$\langle x^2(t) \rangle = \frac{\sigma_1^2}{1-\lambda^2} \tag{10.32}$$

where $\langle \varepsilon^2(t) \rangle = \sigma_1^2 =$ constant if the drift $\langle x(t) \rangle = \lambda^t x(0)$ has been subtracted from $x(t)$. Stationarity is therefore possible if and only if $-1 < \lambda < 1$, where $\lambda = 1$ is called a "unit root" in econometrics. Another macroeconomic way to arrive at (10.31) is simply to regard it as a regression equation and to use standard econometric assumptions to try to test data for its validity. Still a third interpretation is to assert that the monetary authority may try to enforce a rule $x(t) = \lambda x(t-1)$ "to within error" for the next period t, at present time $t-1$ based on the presently observed value $x(t-1)$.

Econometrics and regression analysis aside, from the standpoint of the theory of stochastic processes the model (10.31) with $\lambda = 1$ defines a Martingale process *if the noise $\varepsilon(t)$ is uncorrelated*, and Martingales are inherently nonstationary. That is, stationary noise in (10.31) is impossible if $\lambda = 1$. Indeed, macroeconomists interpret a unit root as necessity to worry about nonstationarity. The simplest Martingale is provided by the Wiener process $B(t)$, but the Wiener process is too simple to describe real markets or macroeconomic noise.

Continuing with regression analysis, suppose instead of (10.31) that we consider a twice time-lagged regression equation

$$x(t) = \alpha x(t-1) + \beta x(t-2) + \varepsilon(t) \tag{10.33}$$

We introduce the time-shift operator $Ly(t) = y(t-1)$. The noise term in equation (10.33) can then be rewritten as equal to

$$x(t) - \alpha x(t-1) - \beta x(t-2) = (1 - \lambda_1 L)(1 - \lambda_2 L)x(t) \tag{10.34}$$

10.7 Integration I(d) and cointegration

Here's the central question in regression analysis: *when is stationary noise possible?* If $-1 < \lambda_1 < 1$ then we can set $\lambda_2 = 0$, but if $\lambda_1 = 1$ then we have

$$(1-L)(1-\lambda_2 L)x(t) = \varepsilon(t) \tag{10.35}$$

or with $x(t,-T) = x(t) - x(t-T)$,

$$x(t,-1) = \lambda_2 x(t,-2) + \varepsilon(t) \tag{10.36}$$

so we see that stationary noise is possible in (10.33) for $-1 < \lambda_2 < 1$ even if equation (10.31) has a unit root. So-called "unit root processes" are central: Martingales describe detrended financial variables.

Ergodic processes are a subclass of stationary processes. Ergodicity means that time averages converge in probability to ensemble averages. By iid is meant statistical independence with stationarity. Iid noise is trivially ergodic, the convergence of time to ensemble averages is provided by the law of large numbers. The stationary process $y(t)$ defined by (10.31) with $|\lambda| < 1$ is ergodic in discrete time: the pair correlations for a time lag nT, $\langle y(t)y(t+nT)\rangle = R(nT)$, vanish as n goes to infinity. This is the sort of ergodicity that's assumed in regression analysis models.[2] With a discrete time stationary process the time average always converges, but if the system is not ergodic then the limit is not necessarily the ensemble average. With a nonstationary process there is no possible appeal to time averages. We know that we must construct ensemble averages in order to perform any data analysis at all.

Inadequate distinction is made in regression analysis between noise levels and noise increments (see Kuersteiner (2002) for an exception). We will clarify this below and will point out that the noise "$\varepsilon(t)$" in the regression equations is always, by necessity, a *noise increment* $\varepsilon(t,-T)$. For example, in equation (10.31) with $\lambda = 1$ the noise is exactly a Martingale increment $x(t,-1) = x(t) - x(t-1)$.

10.7.3 Cointegration

In macroeconomics, relations between economic variables are expected on the basis of non-empirically based equilibrium argumentation. In econometrics, regression analysis is used to try to discover or verify the predicted relationships. Given two time series for two different economic variables x and y, like price levels and the money supply, or FX rates and the relative price levels in two countries, regression analysis assumes a form $y = \alpha + \beta x + \varepsilon(t)$ where the standard assumption in the past was that the noise $\varepsilon(t)$ can be treated as a stationary "error" (nonlinear regression analysis exists in the

---

[2] And is generally absent in the case of Martingales.

literature but is irrelevant here). It was realized well before the rational expectations era that typical macroeconomic variables x and y (price levels and FX rates, for example) are nonstationary. It's also been known since the 1920s that regression analysis based on the assumption of stationary noise can easily "predict" spurious relations between completely unrelated nonstationary variables. The assumption of integration I(d) is that with nonstationary random variables $x(t)$, increments $x(t,T) = x(t + T) - x(t)$ are stationary with T fixed, to within a removable drift. Cointegration was invented as a generalization of the idea of "integration I(d)" as a technique for trying to infer both short-time ($T =$ one period) and long-time equilibrium (based on ergodicity) relations between nonstationary economic variables via regression analysis.

Here's a definition of cointegration quoted literally from Engle and Granger (1987). Think of macroeconomic variables as the components of a column vector $x(t)$. "The components of $x(t)$ are said to be cointegrated of order d,b, denoted CI(d,b), if (i) all components of x are I(d); (ii) there exists a vector $\alpha \neq 0$ so that $z(t) = \hat{\alpha}x$ is I($d - b$), $b > 0$. The vector α is called the co-integrating vector." The "hat" denotes the transpose, a row vector so that $\hat{\alpha}x$ denotes the scalar product of two vectors. The authors then state that for the case where $d = b = 1$, cointegration would mean that if the components of $x(t)$ were all I(1), then the equilibrium error would be I(0), and $z(t)$ will rarely drift far from zero if it has zero mean, and $z(t)$ will often cross the zero line. That is, $\hat{\alpha}x = 0$ is interpreted as an *equilibrium* relationship, and the last part of the sentence above expresses the unproven hope that the stationarity of integration I(d) is the strong stationarity that brings with it the ergodicity of statistical equilibrium (". . . will rarely drift far from zero . . . will often cross the zero line").

The Nobel Committee's description of Granger's work noted that cointegration had failed to exhibit the expected long-time equilibrium relationship expected between FX rates and relative price levels in two countries. It was argued therein that cointegration deals with short times, $T =$ one period, and that short time lags are inadequate to expose the long-time equilibrium relations that would follow from ergodicity. We will show that the real reason for the failure of an equilibrium relation between two financial variables is entirely different, and is not at all due to the restriction to a short time lag T.

Statisticians have constructed simple models where cointegration works, and the conditions to be satisfied are quite restrictive. It's doubtful that two empirical time series will satisfy the conditions for the required noise differences. We therefore challenge the idea that cointegration can be used to explain macroeconomic phenomena. In particular, if (as is always the case in economics) the noise has been postulated rather than discovered empirically,

10.7 Integration I(d) and cointegration

we see the following danger: if one takes enough differences, using sparse data, then spurious "I(d)" and cointegration may occur. That is, the effect of finite, low precision must be taken into account in reality.

In the next two sections we analyze the statistical properties of noise levels and increments. Below, we will show that arbitrary stationary increment Martingales are the right generalization of "white" and iid noise, and that the assumption of iid is both unnecessary and too limiting; lack of increment correlations rather than full statistical independence of increments is adequate. This is also practical, because in empirical analysis we generally cannot discover even a one-point distribution (although various averages and correlations can be calculated empirically), much less prove that hypothetical empirical data are iid even if they were.

10.7.4 The distributions of iid and "white noise" processes and increments

10.7.4.1 Iid noise

Because econometrics and macroeconomics typically assume either "white" or iid noise, we now analyze the statistical properties of both. We will show that an iid (drift-free, statistically independent, identically distributed) noise process $\varepsilon(t)$ generally cannot generate stationary increments, and therefore is not I(d). To avoid confusion, we'll distinguish carefully between the distributions of noise levels and noise level differences. We want to show that the noise "$\varepsilon(t)$" in regression analysis is always a noise increment, and that white noise, not iid noise, is the correct basis for relaxing the restrictions imposed in regression analysis.

Consider an identically distributed statistically independent process, the definition of n identical statistically independent variables $x_1, \ldots, x_n$. All n-point densities of random variables factor into products of the same one-point density $f_n(x_n, t_n, \ldots, x_n, t_n) = f_1(x_n, t_n) \ldots f_1(x_1, t_1)$, $n = 2, 3, \ldots$ An iid distribution is in addition stationary, requiring that f_1 is independent of t.

Consider a nonstationary process $x(t) = \varepsilon(t)$ defined as drift-free, statistically independent, identically distributed noise. For example, the one-point density may be Gaussian with the variance linear in the time t. Here, the increment autocorrelations do not vanish,

$$\langle \varepsilon(t,T)\varepsilon(t,-T)\rangle = -\langle \varepsilon^2(t)\rangle 0 \tag{10.37}$$

For this process a condition of stationarity of increments is impossible: the mean square fluctuation

$$\langle \varepsilon^2(t,T) \rangle = \langle \varepsilon^2(t+T) \rangle + \langle \varepsilon^2(t) \rangle \qquad (10.38)$$

cannot be made independent of t unless the process itself is stationary, but macroeconomic processes are not stationary, and the most fundamental nonstationarity is due to the nature of the noise, not the presence of drift.

One can see the impossibility of stationary increments $\varepsilon(t,T)$ directly by assuming statistical independence in the two-point density while calculating the increment density. With $z = x(t,T) = \varepsilon(t,T)$, we use

$$f(z,t,t+T) = \int_{-\infty}^{\infty} dxdy f_2(y,t+T;x,t)\delta(z-y+x) \qquad (10.39)$$

to obtain

$$f(z,t,t+T) = \int_{-\infty}^{\infty} dx f_1(x+z,t+T) f_1(x,t) \qquad (10.40)$$

If one assumes a nonstationary Gaussian one-point density with process variance $\sigma^2(t)$, then one sees easily that $f(z,t,T)$ depends irreducibly on both T and t. *Stationary increments cannot be achieved under conditions of statistical independence of the stochastic process.* Generally, by "iid" econometricians and statisticians implicitly presume a stationary process; the variance is then constant.

In contrast, for the Wiener process $B(t)$, the simplest Martingale with stationary increments, one obtains from (10.10) that $f(z,t,t+T) = p_2(z,T|0,0)$ where p_2 is the transition density of the Wiener process, $f_2(y,t+T;x,t) = p_2(y,t+T|x,t)f_1(x,t)$. Although the Wiener process $B(t)$ is not iid, the Wiener *increments* $B(t,T)$ are iid in the following precise sense *if and only if* we restrict our considerations to $T =$ constant: the Wiener process $B(t)$ is Markovian,

$$f_n(x_n,t_n;x_{n-1},t_{n-1};\ldots;x_1,t_1) = p_2(x_n,t_n|x_{n-1},t_{n-1})$$
$$p_2(x_{n-1},t_{n-1}|x_{n-2},t_{n-2})\ldots$$
$$p_2(x_2,t_2|x_1,t_1)p_2(x_1,t_1|0,0) \qquad (10.41)$$

If we combine the Markov condition with the time- and space-translational invariance of the Wiener process $p_2(x_n,t_n|x_{n-1},t_{n-1}) = p_2(x_n - x_{n-1}, t_n - t_{n-1}|0,0)$, then this casts (10.41) into the form of a condition for iid *increments* if we take $t_k - t_{k-1} = T$ for all k, where the one-point density of increments is exactly $f(z,t,t+T) = p_2(z,T|0,0)$.[3] That is, the random walk

---

[3] Without both time- and space-translational invariance, one cannot obtain an iid distribution for increments from a Markov condition.

10.7 Integration I(d) and cointegration

is not iid, but random walk increments are iid. Note that with T varying, the increment process is nonstationary, but with T = constant we may treat random walk increments as iid. *The basic example of iid increments is not a stationary process, and the iid property is simply time- and space-translational invariance applied to a Markov condition.* This is unnecessarily restrictive. We don't need the iid assumption; we only need uncorrelated increments.

Again, for I(1) noise $x(t)$, all we need is that $x(t,1)$ and $x(0,1)$ have the same *one-point* distribution. *Full statistical independence is unnecessary; all we need is the condition of uncorrelated, stationary increments.* Second, deducing iid conditions from empirical data would be effectively impossible; the best that can be hoped for empirically is to test for vanishing increment autocorrelations. Therefore, the iid assumption can and should be replaced by the much more general condition of a Martingale with stationary increments (Martingales are Markovian if and only if there is no finite memory in the transition density). We will eventually have to face and answer the question: what class of Martingales has stationary increments? First, more background.

10.7.4.2 "White noise" in econometrics

To support our claim of the importance of Martingales for integration I(d), here's the simplest I(1) case presented in the literature. Consider the random walk on the line, the Wiener process $B(t)$. The Wiener process is a Martingale, $\langle B(t)B(t-T)\rangle = \langle B^2(t-T)\rangle$, with stationary increments: $B(t,T) = B(t+T) - B(t) = B(T)$ in distribution because the variance is linear in t, $\langle B^2(t)\rangle = t$. The increments are therefore uncorrelated, $\langle B(t,T)B(t,-T)\rangle = -\langle B(T)B(-T)\rangle = 0$. That is, the process $B(t)$ has Martingale pair correlations and the increments are uncorrelated and stationary. The increment viewed as a *process* $B(t,T) = B(0,T) = B(T)$, with $B(0) = 0$, is therefore nonstationary in T because the increment variance is linear in T. Economists study increments with a fixed time lag T = one period.

The Wiener increment process has been labeled as "white noise" in econometrics (Murray, 1994). In the economists' "white noise," the increments are uncorrelated, $\langle B(t,1)B(t,-1)\rangle = 0$, and have constant variance $\langle B^2(1)\rangle = 1$, nothing more. In other words, "white noise" in econometrics is actually the simplest Martingale with stationary increments and fixed time lag. *We can therefore de-emphasize "white noise" and iid noise and focus instead on the much more general case of stationary increment Martingales in order to define integration I(d) for drift-free stochastic processes* (McCauley, 2009).

With T fixed, any stationary increment Martingale is I(1). For Martingales, I(d) with $d \geq 2$ is superfluous.

10.7.5 Integration I(d) via stationary increment martingales

An Ito process is generated locally by a drift term plus a Martingale $dx = b(x,t)\,dB$. Setting $b^2(x,t) = D(x,t) = D(t)$, or $b(x,t) = x$ (lognormal process), in an Ito process generates two different Martingales, but each is equivalent to the Wiener process by a specific coordinate transformation (McCauley et al., 2007c). Setting $D(x,t) = |t|^{2H-1}(1+|x|/|t|^H)$ with $H \approx 0.35$ in (7.34) describes a nontrivial Martingale observed during one time interval of intraday FX trading. This Martingale is topologically inequivalent to the Wiener process.

Martingales may serve as noise sources in a generalization of standard regression analysis, but are themselves not subject to regression analysis: a linear combination of two Martingales is another (local) Martingale, as one can see from the Martingale representation theorem. Given two Martingales defined by two independent Wiener processes, the two Martingales are also independent. *Stationary increment Martingales define the class of pure noise processes that are* I(1). Exactly what is the class of stationary increment Martingales?

Consider first the class of all drift-free nonstationary processes with uncorrelated stationary increments. For this class the one-point density of the random variable x is nonstationary, and the increments $x(t,T)$ and $x(0,T)$ have the same nonstationary one-point distribution as a function of the time lag T as has the one-point distribution of the process $x(t)$ as a function of the starting time t.

The simplest example of a Martingale where I(1) is impossible is a Martingale with $b(t)$ independent of x, and where the increments are nonstationary unless $b(t) =$ constant. But this increment nonstationarity can be easily eliminated, yielding an I(1) process, by discovering the time transformation that reduces the process to the Wiener process. The required transformation can easily be constructed once $b(t)$ is known, and $b(t)$ could be discovered *if* one could measure the time dependence of the process variance

$$\langle x^2(t)\rangle = \int_0^t b^2(s)\,ds \qquad (10.42)$$

As an example, $b(t) = t^{H-1}$ yields $\langle x^2(t)\rangle = t^{2H}\langle x^2(1)\rangle$ so the time-transformation $\tau = t^{2H}$ yields stationary increments.

If we consider general diffusive processes, with diffusion coefficient $D(x,t)$, then we obtain a mean square fluctuation independent of t if and only if the variance is linear in t. This occurs for scaling processes $D(x,t) = |t|^{2H-1}\widehat{D}(|x|/|t|^H)$ with $H = 1/2$, but these processes generally do not satisfy the condition for a t-independent increment density $f(z,t,T)$. For time-translationally

10.7 Integration I(d) and cointegration

invariant diffusive processes, $D(x,t) = D(x)$, but these processes have variance nonlinear in t. The simplest example is given by the drift-free lognormal process

$$dp = pdB \tag{10.43}$$

where B is the Wiener process. The first differences are Martingale increments

$$p(t+T) - p(t) + \int_t^{t+T} p(s)dB(s) \tag{10.44}$$

and are nonstationary,

$$\langle (p(t+T) - p(t))^2 \rangle = \int_t^{t+T} \langle p^2(s) \rangle ds \tag{10.45}$$

because $\langle p^2(t) \rangle = Ce^t$. Is there a nontrivial stationary increment Martingale?

With the increment density

$$f(z,t,t+T) = \int dx p_2(x+z, t+T|x,t) f_1(x,t) \tag{10.46}$$

if we assume time-translational invariance, then we also need space-translational invariance to obtain

$$f(z,t,t+T) = p_2(z, t+T|0,t) \int dx f_1(x,t) = p_2(z, t+T|0,t) \tag{10.47}$$

but this implies the Wiener process! *We speculate that the Wiener process is the only stationary increment Martingale.* If true, this means that cointegration is built on a noise model too simple to describe real empirical data. There is therefore no reason to assume that differences obtained from real data can generate ergodicity.

This leads us to the interesting question: what methods have been used to claim stationarity of differences, "integration I(1)," in macroeconomic data analysis (Dickey *et al.*, 1991)? Given the standard statistical tests including the search for a unit root, we suspect that the answer to our question ranges from completely inadequate evidence to none at all. *The unit root test is an insufficient test for a Martingale, and even then does not test for increment stationarity.* The evidence is often provided only by the visual inspection of a time series (see the graphs of levels and differences in Juselius and MacDonald, 2003). A more convincing argument would require using an ensemble average to show that a mean square fluctuation is t-independent.

Where increments can be measured accurately using ensemble averages for intraday finance data, the increments are strongly nonstationary. In any case, from the standpoint of Martingales as financial returns there's no reason to expect a long-time equilibrium based on ergodicity for FX rates and relative price levels for two different countries. The required convergence of time averages to ensemble averages simply does not exist.

10.7.6 Nonstationary increment martingales are not I(d)

Here's the most general case: integration I(d) is in principle impossible, whether for $T =$ one period or for any value of T, if Martingale noise increments are nonstationary, as indeed they are in finance data.

Figure 7.2a shows the nonstationarity of increments in intraday trading as the time variation of the ensemble average of the diffusion coefficient $D(x,t)$. The scatter in Figure 7.2 is due to inadequate intraday statistics caused by the ensemble average required to handle nonstationary increments correctly. Each day can be understood as a rerun of the same uncontrolled experiment if we restrict to certain averages and avoid others, and in a six-year time series, there are only 1500 points for each intraday time t (1500 days for each time t) from which to calculate ensemble averages.

With nonstationary increments, the increment $x(t, -T) = x(t) - x(t - T)$ depends irreducibly on the starting time t, and no amount of higher-order differencing can eliminate this nonstationarity in principle. That is, integration I(d) is impossible for intraday FX data; differencing cannot lead to stationarity for financial time series. A visual inspection of Figure 7.2b, for the ensemble average result for the rmsf over a week, indicates visually that the increments "look" approximately stationary on the time scale of a day. But if one would try to verify that, then the histograms would have 1500 points for one day, 750 points for the second day; far too few to reach a conclusion.

The longest line in Figure 7.2a shows the region for which we were able to fit the FX data via a scaling Martingale,

$$D(x,t) = |t|^{2H-1}(1 + |x|/|t|^H)/H \qquad (10.48)$$

with $H \approx 0.35$, and this model explicitly has nonstationary increments with $H \neq 1/2$. The fit is shown as Figure 10.1.

What about loopholes? Is it possible to "flatten" the mean square fluctuation of the Martingale describing intraday trading by a time transformation? For the model (10.49) a time transformation $\tau = t^{2H}$ yields a variance linear in τ, but this model only describes trading over one small segment of a day. Using a two-year Olsen and Associates FX series, Gallucio et al. (1997) performed a local time transformation to obtain a time-independent mean

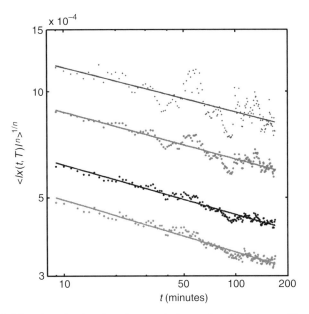

Figure 10.1 The data in the time interval shown by the longest line in Figure 7.2a can be fit by scaling with a single Hurst exponent $H \approx 0.35$.

square fluctuation. This might be possible piecewise numerically using a transformation to a local lag time τ, $t \ll T$, where (analogous to the procedure described by Gallucio *et al.*)

$$\tau \approx \int_{t}^{t+T} \langle x^2(s,T) \rangle ds \tag{10.49}$$

and where

$$\langle x^2(t,T) \rangle \approx T \int dx D(x,t) f_1(x,t) \tag{10.50}$$

describes Figure 10.7, but we haven't bothered to try this recipe numerically. We would not be able to construct such a transformation globally (over a day or longer) accurately due to too much scatter in the fit of (10.50) to the data. And even if we could succeed, ergodicity would not be part of the package. Nor could we conclude that the increment density is *y*-independent; *t*-independence of the rmsf is too weak a condition for that purpose. This brings us to the next question.

One may ask if tick data (Jensen *et al.*, 2004; Politi and Scalas, 2008) yield a flat rmsf. A moment's reflection shows that the transformation from tick time to real time is not deterministic: given a tick, we cannot state with certainty when the next tick occurs, so we cannot transform systematically from a real-time analysis to a tick analysis. If the ticks are recorded in real time then a local time transformation from ticks to real time would be numerically possible. Recording ticks in real time, the nonstationarity of the increments can be understood (Cross (1988)) in terms of lunch breaks, the closing time of New York markets, the closing of one international market, and the opening of another. This is an indication that intraday increments may have a flat rmsf in tick time, but this doesn't help us with either cointegration or with real-time data analysis.

10.8 ARCH and GARCH models of volatility

In discussions of ARCH and GARCH models, as in regression analysis in general, an inadequate distinction is made between noise levels and noise increments as we've discussed above. The noise in regression equations must be interpreted as noise increments, whether one assumes iid or "white noise." Therefore, what Engle originally called a "variance" should be understood as a mean square fluctuation, or "increment variance." We've shown above that "white noise" in econometrics means stationary noise increments with vanishing increment autocorrelations.

There are various volatility measures in practical use in finance theory. The volatility measure chosen by Engle is the conditional mean square fluctuation $V(t,T) = \langle x^2(t,-T) \rangle_{cond}$. In a diffusive model (an Ito process) this would be given by

$$\langle x^2(t,-T) \rangle_{cond} = \int dy (y-x)^2 p_2(y, t+T|x, t) \tag{10.51}$$

where p_2 is the conditional density for the returns process $x(t)$.

In all that follows, we assume detrended data and/or detrended stochastic models of levels $x(t)$. This severely restricts the class of models to those where the drift is neither a function nor a functional of x. With the choice $x(0) = 0$ the process variance is given by $\sigma^2(t) = \langle x^2(t) \rangle$, where the process $x(t)$ is then drift-free noise. Only uncorrelated noise increments are of interest here, the time lag T must be sufficient that $\langle x(t,T)x(t,-T) \rangle \approx 0$, ruling out fBm and other strongly correlated stationary increment processes. Next, we consider the basic regression models of volatility.

The standard statement of an ARCH(1) process is that with $\varepsilon_t / \langle y_t^2 \rangle$ assumed to be white noise, then

$$\langle y_t^2 \rangle_{cond} = \alpha + \omega y_{t-1}^2 \tag{10.52}$$

where the detrended returns are described by $\varepsilon_t = \ln(p(t)/p(t-1))$. Clearly, as has been pointed out recently, both the noise and the variable y here are

10.8 ARCH and GARCH models of volatility

not levels; they are both increments. Having made this point, we now return to our standard notation for increments. We've pointed out elsewhere that it's quite common, if mistaken, to regard the log increment $x(t) = \ln(p(t)/p(t-T))$ as a process, or level.

Historically, ARCH models were introduced to remedy the lack of volatility of the Gaussian returns model in the 1980s. The ARCH models were constructed with memory intentionally built into the mean square fluctuation. Whether or not it was realized that the EMH is violated is not clear, because previous discussions of Martingales as the EMH focused on simple averages and either ignored pair correlations or stated them incorrectly (see Chapter 4). The contradiction between ARCH and the EMH was probably masked by failing to distinguish between levels and differences in the noise, and by taking $T = 1$ instead of letting T vary.

The ARCH(1) model is defined by the regression equation

$$\langle x^2(t, -T)\rangle_{\text{cond}} = \alpha + \omega x^2(t-T, -T) \tag{10.53}$$

with the assumption that the increments are stationary, and hence are independent of t. In addition, the assumption was made that

$$x(t, -T) = z(T)\langle x^2(t, -T)\rangle_{\text{cond}}^{1/2} \tag{10.54}$$

where $z(T)$ was originally taken to be iid with zero mean and unit variance. It's adequate to assume that $z(T)$ is uncorrelated with zero mean and unit variance. The idea is that $x(t,-T) = x(0,-T)$ "in distribution" is the stationary noise in regression equations (10.53) if T is held fixed. So far, this is completely in the spirit of regression analysis: the noise is not assumed to have been discovered empirically; it's postulated in as simple a way as possible.

The unconditioned averages in ARCH(1) then obey

$$\langle x^2(t, -T)\rangle = \alpha + \omega\langle x^2(t-T, -T)\rangle \tag{10.55}$$

In regression analysis the assumption is that the increments are stationary. Stationary increments have been hypothesized on the basis of "eyeballing" plots of levels and differences, but were never verified by a statistical analysis based on constructing approximate ensemble averages. Accepting the assumption of stationary increments for the time being, we obtain

$$\langle x^2(t, -T)\rangle = \langle x^2(t-T, -T)\rangle = \langle x^2(0, -T)\rangle \tag{10.56}$$

independent of t. This would yield

$$\langle x^2(0, -T)\rangle = \frac{\alpha(T)}{1 - \omega(T)} \tag{10.57}$$

This is a T-dependent relationship that could be checked, but that fact is masked by setting $T = 1$ in regression analysis. We now show, without appeal to any particular dynamics, that the ARCH(1) model is completely inconsistent with "white noise" (uncorrelated noise differences).

Increment autocorrelations are given by

$$2\langle x(t,-T)x(t,T)\rangle = \langle (x(t+T)-x(t-T))^2\rangle - \langle x^2(t,-T)\rangle - \langle x^2(t,T)\rangle \quad (10.58)$$

With stationary increments we obtain

$$2\langle x(0,-T)x(0,T)\rangle = \langle (x(0,2T))^2\rangle - 2\langle x^2(0,T)\rangle \quad (10.59)$$

The increment autocorrelations vanish if and only if the levels' variance is linear in the time, which then yields also that $\langle x^2(0,T)\rangle = T\langle x^2(0,1)\rangle$. Inserting this into (10.66), if we set $T = 0$ then we obtain $\alpha = 0$. If $T \neq 0$ then we obtain $\omega = 0$ (regression analysis therefore fails). *This shows that ARCH(1) is inconsistent with stationary, uncorrelated increments.* The same conclusion will hold if the increments are nonstationary and uncorrelated. The reason for the contradiction is clear: uncorrelated increments guarantee a Martingale $x(t)$, and the Martingale condition rules out memory at the level of simple averages and pair correlations. ARCH models, in stark contrast, have finite memory built in at the pair correlation level. The correct way to understand the ARCH models is that the memory requires nonvanishing increment correlations. This violates the EMH and finance data as well. Higher order ARCH models admit exactly the same interpretation.

The GARCH(1,1) model is defined by

$$\langle x^2(t,-T)\rangle_{\text{cond}} = \alpha + \omega x^2(t-T,-T) + \zeta \langle x^2(t-T,-T)\rangle_{\text{cond}} \quad (10.60)$$

If we again assume stationary increments then we obtain an analogous constant mean square fluctuation for fixed T. In this case "white noise" would imply that $\alpha = 0$ and that $\omega + \zeta = 0$. With enough parameters the models are not falsifiable. There is no evidence for memory in observed finance market returns for $T \geq 10$ minutes. ARCH and GARCH models are only applicable to processes with correlated increments, and not to "white noise" processes. In financial applications this requires lag time of $T < 10$ minutes in trading. Correlated increments characterize fBm, while uncorrelated increments characterize an efficient market.

Summarizing, no existing regression model (cointegration, ARCH/GARCH) describes finance data qualitatively correctly because the empirical conditions necessary to apply those ideas are not met.

11
Complexity

11.1 Reductionism and holism

It was a fad in certain circles in the 1990s to announce that "reductionism is dead," and that holism[1] is necessary in order to understand biology and social systems. To discuss this we must first define the terms (The American Heritage Dictionary).

Holism is a theory or belief emphasizing the importance of the whole and the interdependence of its parts. In classical mechanics, for example, we understand the solar system in terms of its parts (the sun and the planets), and the interactions ("relationships") are nonlinear. In quantum mechanics we understand complicated molecules like DNA in terms of the binding of smaller molecules. The genetic code is understood in terms of its building blocks: the four letter (computer) alphabet constructed from the bases A, C, G, and T. One understands the bases in terms of simpler molecules, the molecules in terms of atoms, and the atoms in terms of electrons and nuclei. This is reductionism: everything can be understood to be constructed systematically from electrons and positively charged nuclei. Yet, there is something that we cannot calculate from first principles using quantum mechanics: the behavior or the genes as a classical computer with three-letter words constructed from a four-letter alphabet. Quantum theory is linear, and the phase coherence cannot be destroyed from within, meaning that classical behavior cannot be derived from quantum theory without introducing a nonquantum assumption of destruction of phase coherence. This leads into the famous measurement problem clarified by von Neumann. The genetic code and protein production are understood via reductionism. In fact, every important discovery in biology, from the chromosomes to the genetic code to cancer-causing mutations are results of reductionism, which we can understand in

---

[1] Holism was thought to provide a pathway to understand complexity.

a general way as the isolation of cause and effect. This is obviously not what people mean when they claim that reductionism has failed, and that holism offers new hope.

If we visit Wikipedia (http://en.wikipedia.org/wiki/Holism) then we find a definition of holism more to the liking of those who claim that reductionism has failed us: holism ... is the idea that *not* all the properties of a given system (ruling out physical systems like genes and protein production) can be explained by its components alone. Instead, the system as a whole determines in an important way how the parts behave. Actually, the fact that not all properties of a system can be understood in terms of components is *not* "holism"; there may well be properties that cannot be understood at all (like consciousness and thinking). This idea of holism is prescientific and was expounded by Aristotle, who wrote that "the whole is more than the sum of its parts." If we would interpret the term "sum" literally, then all that's stated is that the system is nonlinear (which is certainly *not* true of quantum systems like DNA). But this leads us trivially back to physics as the basis for understanding everything. Indeed, the three-body problem in classical mechanics, and hence the solar system as well, is irreducibly nonlinear (in contrast, the also nonlinear sun–earth two-body problem can be linearized exactly by a coordinate transformation). As Galileo discovered, Aristotle is not a reliable guide to scientific advances; he had to be completely sidestepped by Archimedes, Galileo, and Kepler before science could begin and then advance. This doesn't prevent some mysticists from believing that Aristotle was right.

Some earlier system biologists have argued that nonlinear systems are holistic. That argument is flawed: *anything that can be mathematized is an example of reductionism*, meaning we can understand the system in terms of its parts, the parts being represented by the terms in the nonlinear equation. What Aristotle had in mind was nonscientific, namely, a "first cause," which Tomaso d'Aquino replaced by the idea of God. But the idea of a god who tweaks the universe to keep it stable, or who tweaks molecules to make life, is an unnecessary assumption.

Reductionism is isolation of cause and effect. Let us accept that reductionism is the understanding of the whole in terms of its parts. Thus do we understand the solar system and the building of proteins from DNA. Strict reductionism would assert that, in principle, we "understand" everything once we've understood quarks. Or, we understand almost everything in principle once we've understood electrons and nuclei and how they interact. This is not a useful viewpoint. In fact, it's downright silly. We know that electrons and nuclei make up the atoms in a turbulent flow, but only an uninformed greenhorn would suggest that we try to understand fluid

11.1 Reductionism and holism

turbulence by starting with electrons and protons. We don't even understand fluid turbulence in mathematical detail from the Navier–Stokes equations. Reductionism doesn't mean that we can calculate everything from electrons and nuclei, or from quarks. As Rolf Landauer once stated, reductionism means that we can successfully divide the world into "system plus environment" and can, to zeroth order, neglect the interaction of the system with the environment. If we can't do that then we can't study "the system" at all. To understand cellular interactions we try to start with cells as basic units, cells being well-defined and persistent if not invariant, but we certainly cannot completely ignore what happens inside a cell. On the other hand, medical doctors could not, for example, treat liver disease were it always necessary to take into account the brain, the heart, the nervous systems, etc., because no one can make any sense of such complexity. It would be worse than trying to repair a motor or radio by trying to think of all the parts and their interconnections simultaneously. All of scientific medicine is reductionist. "Holistic medicine" rears its head where no treatment is known, and never offers any systematically verifiable solution, only false hope for uncritical believers. Some herbal remedies systematically work, and there are scientific reasons why.

Economic models ignore nearly *everything* in the environment in the effort to try to get a handle on *something*. The attempt to derive a mathematical model from a time series in finance is an example of reductionism. The component parts are the prices. At a deeper level, researchers try to invent agent-based models. There, the components are rules representing idealized agents who try to set prices. Generally, it makes no sense whatsoever to try to look into the minds of agents because either (i) the effort is destroyed in advance by nonuniqueness (due to finite precision; see Chapter 7) or else (ii) trivialized by unrealistic assumptions like utility and price preferences in neo-classical economics. In contrast with the failure of microeconomic theory, our path is inherently nonstandard macroeconomic. We've used historic market price series to try to understand markets: we extracted a correct class of price dynamics models from observed time series, Martingales with (x,t)-dependent diffusion coefficients.

Reductionism means the arbitrary division of nature into laws of motion and initial conditions, plus "the environment." The initial conditions are lawless, are not themselves derived from an identifiable equation of motion. The division into laws of motion and approximately *uncorrelated* initial conditions is very important for analyzing finance time series via an approximate statistical ensemble in Chapter 7. There, we replaced Wigner's assertion that initial conditions should be "random" by the less restrictive and more directly applicable requirement that initial conditions must be uncorrelated, otherwise they are not initial conditions but effectively are determined by a law of motion. This summarizes our viewpoint on reductionism, but the question of complexity has not been addressed yet. The failed hope was that holism would explain or at least handle complex systems.

11.2 What does "complex" mean?

Is market behavior complex? What's meant by "complexity"? In particular, does the word have definite dynamical meaning? How does complexity differ from complication? From chaos? From randomness? Can scaling be used to describe complexity? Because the word "complexity" is most often used without having been explicitly and clearly defined, we delineate what is complex from what is not on the basis of standard ideas of computational complexity.

The reason for this choice of reference is that we still lack a convincing physically or biologically motivated definition of complexity, in spite of the fact that cell biology apparently provides us with plenty of examples of complexity. A digital computer provides an example of complexity and can be described mathematically as a Newtonian electro-mechanical machine. The only precise definitions of complexity that have been used so far in physics, biology, and nonlinear dynamics are definitions that were either taken from or are dependent on computability theory.

The first idea of complexity to emerge historically was that of the highest degree, equivalent to a Turing machine. Ideas of degrees of complexity, like how to describe the different levels of difficulty of computations or how to distinguish different levels of complexity of formal languages generated by automata, came later.

We begin with binary strings because from a fundamental standpoint there's nothing computable that can't be encoded as a binary string, or as a sequence of strings (whether binary, ternary, or decimal is a detail). As von Neumann (1970a) wrote, decimal expansions are an application of information theory. Digit strings can be regarded as patterns.

A systematically repeated pattern in finance data would violate the EMH and could in principle be exploited to make unusual profits in trading. By an unusual profit we mean a profit greater than the expected return as discussed in CAPM. We could search for patterns in economic data as follows: suppose that we know market data to three-decimal accuracy; for example, after rescaling all prices p by the highest price so that $0 \leq p \leq 1$. This would allow us to construct three separate coarsegrainings: empirical histograms based on ten bins, 100 bins, and 1000 bins. Of course, because the last digit obtained empirically is the least trustworthy, we should expect the finest coarsegraining to be the least reliable one. In the ten-coarsegraining each bin is labeled by one digit (0 through 9) while in the 1000 coarsegraining each bin is labeled by a triplet of digits (000 through 999). An example of a pattern would be to record the time-sequence of visitation of the bins by the market in a given coarsegraining. That observation would produce a sequence of digits, called a symbol sequence. The question for market analysis is whether a pattern systematically nearly repeats itself. Mathematically well-defined symbolic dynamics is a signature of deterministic chaos, or of a deterministic dynamical system at the transition to chaos.

First, we present some elementary number theory as the necessary background. We can restrict to numbers between zero and unity because, with those numbers expressed as digit expansions (in binary, or ternary, or in any base of arithmetic) all possible one-dimensional patterns that can be defined to exist abstractly exist there. Likewise, all possible two-dimensional patterns arise as digit expansions of pairs of numbers representing points in the unit square, and so on. Note that by "pattern" we do not restrict ourselves to a periodic sequence; nonperiodic sequences are included.

We can use any integer base of arithmetic to perform calculations and construct histograms. In base μ we use the digits $\varepsilon_k = 0, 1, 2, \ldots, \mu-1$ to represent any integer x as $x = \sum \varepsilon_k \mu^k$. In base ten the digit 9 is represented by 9, whereas in base two the digit 9 is represented by 101, and in base three 9 is represented by 22. Likewise, a number between zero and one is represented by $x = \sum \varepsilon_k \mu^{-k}$. We will mainly use binary expansions ($\mu = 2$) of numbers in the unit interval in what follows, because all possible binary strings/patterns are included in that case. From the standpoint of arithmetic we could as well use ternary, or any other base.

Finite-length binary strings like 0.1001101 (meaning 0.100110100000000... with the infinite string of 0s omitted) represent rational numbers that can be written as a finite sum of powers of 2^{-n}, like $9/16 = 1/2 + 1/2^4$. Periodic strings of infinite length represent rational numbers that are not a finite sum of powers of 2^{-n}, like the number $1/3 = 0.010101010101\ldots$, and vice versa. Nonperiodic digit strings of infinite length represent irrational numbers, and vice versa (Niven, 1956). For example, $\sqrt{2} - 1 = 0.0110101000001001\ldots$ This irrational number can be computed to as high a digital accuracy as one pleases by the standard grade school algorithm.

We also know that every number in the unit interval can be formally represented by a continued fraction expansion. However, to use a continued fraction expansion to generate a particular number, we must first know the initial condition or "seed." As a simple example, one can solve for the square root of any integer easily via a continued fraction formulation: with $\sqrt{3} = 1+x$, so that $0 < x < 1$, we have the continued fraction $x = 2/(2 + x)$. In this formula the digit 2 in the denominator is the seed (initial condition) that allows us to iterate the continued fraction, $x = 2/(2 + 2/(2 + \ldots))$ and thereby to construct a series of rational approximations whereby we can compute $x = \sqrt{3} - 1$ to any desired degree of decimal accuracy. Turing (1936) proved via an application of Cantor's diagonal argument (Hopkin and Moss, 1976) that for almost all numbers that can be defined to "exist" abstractly in the mathematical continuum there is no seed: almost all numbers (with measure one) that can be defined to exist in the mathematical continuum are both irrational and not computable via any possible algorithm. The measure-zero set of irrational numbers that have an initial condition for a continued fraction expansion were called computable by Turing. Another way to say it is that Turing proved that the set of all algorithms is countable, is in one-to-one correspondence with the integers. This takes us to the original idea of maximum computational complexity at the level of the Turing machine.

11.2.1 Computable numbers and functions

Mathematics is required to describe theoretical mechanics, but arithmetic can be understood as a mechanical operation: Alan Turing mechanized the idea of computation systematically by defining the Turing machine. A Turing machine can in principle be used to compute any computable number or function (Turing, 1936). We can recursively construct a computable number or function, digit by digit, using only integers in an algorithm. The algorithm can be used to generate as many digits as one wants, within the limits set only by computer time. Examples are the continued fraction expansion for $\sqrt{2}$ and the grade school algorithm for $\sqrt{2}$.

An example of recursion is the logistic map $x_n = Dx_{n-1}(1 - x_{n-1})$ with control parameter D. Recursion alone doesn't guarantee computability: if the initial condition x_0 is noncomputable, or if D is noncomputable, then so are all of the iterates x_n for $n > 0$. If, however, we choose as initial condition a computable number like $x_0 = \sqrt{2} - 1$, and a computable control parameter like $D = 4$, then by expressing both the initial condition and the map using binary expansions $x_n = 0.\varepsilon_1(n)\ldots\varepsilon_N(n)\ldots$, where $D = 4 = 100$ in binary, the logistic map defines a simple automaton/machine from which each point of the orbit $x_0, x_1, \ldots, x_n, \ldots$ can be calculated to as many decimals as one wants, always within the limits set by computation time (McCauley, 1993, 1997a). Information is lost only if one truncates or rounds an iterate, but such mistakes are unnecessary (in grade school, such mistakes are penalized by bad grades, whereas scientific journals during the last 25 years have typically rewarded them). We have just described an example of an exact, computable chaotic trajectory calculated with controlled precision.

A noncomputable number or function is a number or function that cannot be algorithmically generated digit by digit. No one can give an example of a noncomputable number, although such numbers "fill up" the continuum (are of measure one). If we could construct market statistics by a deterministic model or game, then market statistics would be algorithmically generated. This would not necessarily mean that the model or game is complex. But what is the criterion for complexity? Let's survey next a useful attempt to define complexity.

11.2.2 Algorithmic complexity

Consider a binary string/pattern of length n. The definition of the algorithmic complexity of the string is the length K_n of the shortest computer program that can generate the string. The algorithm is the computer program. To keep the discussion focused, let us assume that machine language is used on a binary computer. The longest program of interest is: to write the digits one after the other, in which case $K_n = n$.

The typical sort of example given in popular papers on algorithmic information theory is that 101010101010 should be less complex than a nonperiodic string like 100100011001. From a naive standpoint both strings would appear to be equally simple. For example, seen as binary fractions, $0.1010 = 5/8$ whereas $0.1001 = 9/16$. Every finite binary string can be understood as either a binary fraction or an integer ($1010.0 = 5$ and $1001.0 = 17$, for example). Instead of writing the string explicitly, we can state the rule for any string of finite length as: write the binary expansion of the integer or divide two integers in binary. All rational numbers between zero and unity are specified by an algorithm that states: divide integer P by integer Q. But the number of steps required to carry out these divisions grows in length as the denominator grows in size. Intuition is dangerous here. Stefano Zambelli has programmed Turing machines on his laptop to show that the Kolmogorov complexity of various simple-looking binary sequences can differ by orders of magnitude (Velupillai, 2005a). One can prove that almost all numbers (in the sense of measure one), written as digit expansions in any integer basis of arithmetic, are algorithmically complex (Martin-Löf, 1966).

We can summarize by saying that many periodic binary sequences are simple, and that some nonperiodic strings are also simple because the required algorithm is short, like computing $\sqrt{2}$. From this perspective, nonperiodic computable sequences that are constructed from irreducibly very long algorithms are supposed to be more complex, and these sequences can be approximated by rational sequences of long period. Unfortunately, this definition still does not give us any "feeling" for or insight into what complexity really means physically, economically, or biologically. Also, the shortest algorithm that generates a given sequence may not be the one that nature (or the market) uses. For example, one can generate pictures of mountain landscapes via simple algorithms for self-affine fractals, but those algorithms are not derived from physics or geology and in addition provide no insight whatsoever into how mountains actually are formed, showing that the shortest algorithm doesn't necessarily explain the phenomenon physically.

What about the idea of complexity from both simple seeds *and* simple algorithms? The logistic map is not complex but generates chaotic orbits from simple binary initial conditions, like $x_0 = 1/8$. That is, the chaos is "manufactured" from simplicity ($1/8 = 0.001$) by a very simple algorithm. Likewise, we know that there are one-dimensional cellular automata that are equivalent to a Turing machine (Wolfram, 1983, 1984). However, the simpler the machine, the more complicated the program. There is apparently no way to get complexity from simple dynamics plus a simple initial condition.

11.2.3 Automata

Can every mathematical problem that is properly defined be solved? Motivated by this challenging question posed by Hilbert, Turing (1936) mechanized the idea of computation and generalized the notion of typing onto a ribbon of unlimited length to define precisely the idea of a universal computer, or Turing machine. The machine is capable of computing any computable number or function and is a formal abstraction of a real, finite computer. A Turing machine has unlimited memory. By proving that almost all numbers that can be defined to exist are noncomputable, Turing proved that there exist mathematical questions that can be formulated but not definitively answered. For example, one can construct computer programs that do not terminate in finite time to yield a definite answer, representing formally undecidable questions.

Von Neumann (1970a) formalized the idea of abstract mechanical systems, called automata, that can be used to compute. This led to a more useful and graphic idea of abstract computers with different degrees of computational capability. A so-called "universal computer" or universal automaton is any abstract mechanical system that can be proven to be equivalent to a Turing machine. The emphasis here is on the word *mechanical*, in the sense of

11.2 What does "complex" mean?

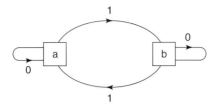

Figure 11.1 The two-state automaton that generates the Thue-Morse sequence.

classical mechanical: there is no randomness in the machine itself, although we can imagine the use of random programs in a deterministic machine. One can generate a random program by hooking a computer up to radioactive decays or radio noise, for example.

In thinking of a computer as an automaton, the automaton is the dynamical system and the program is the initial condition. A universal binary computer accepts all possible binary programs. Here's an example of a very simple automaton, one that is far from universal: it accepts only two different programs and can compute only very limited results. Starting with the binary alphabet $\{a, b\}$ and the rule R whereby a is replaced by ab and b by ba we can generate the nonperiodic sequence $a, ab, abba, abbabaab, abbabaabbaababba, \ldots$ The finite automaton in Figure 11.1 computes the Thue–Morse sequence in the following way. Consider the sequence of programs 0, 1, 10, 11, 100, 101, 110, 111, 1000, ... to be run sequentially. Before running each separate program, we agree to reset the machine in the state a. The result of all computations is recorded as the combined sequence of outputs for each input, yielding the Thue–Morse sequence: abbabaabbaababba... Note that the machine simply counts the number of as in each program mod 2, and that the separate programs are the integers 0, 1, 2, 3, ... written in base two.

Addition can be performed on a finite automaton but multiplication, which requires increasing the precision (increasing the number of bits held in the registers and output) rapidly during the calculation, requires an automaton of unlimited size (Hopkin and Moss, 1976). Likewise, deterministic chaos requires increasing the precision within which the initial condition is specified at a rate determined by the largest Liapunov exponent λ. For an iterated map $x_n = f(x_{n-1})$ with $\lambda = \ln 2$, for example, we must increase the number of bits specified in the initial condition x_0 (written as a binary string) at the rate of one bit per iteration of the map. As an example, if we choose $x_0 = 1/8$ for the logistic map $x_n = 4x_n(1 - x_{n-1})$ and write all numbers in binary ($4 = 100$, for example), then we obtain the orbit $x_0 = 0.001$, $x_1 = 0.0111$, $x_2 = 0.111111$, $x_3 = 0.0000111111, \ldots$ The effect of the Liapunov exponent $D = 4 = e^{\ln\lambda} = 100$ is to shift the third bit of the simple product $x_{n-1}(1 - x_{n-1})$ into the first bit of x_n, and also to tell us the rate at which we must expect to increase the precision of our calculation per iteration in order to avoid making a mistake that eventually will be propagated into an error in the first bit. This orbit is chaotic but it is neither random (it is pseudorandom) nor is it complex: the required algorithm is simple. The level of machine complexity required for computing deterministic chaos here is simply the level of complexity required for multiplication, plus adequate memory for storing digit strings that grow in length at the rate $N_n \approx 2^n N_0$ where N_0 is the number of bits in the initial condition (McCauley, 1993).

How do we know when we have a complex pattern or when we have complex dynamics? In the absence of a physically or biologically motivated definition of degrees of complexity, we must rely on definitions of levels of complexity in computer science, like NP-completeness (Hopcroft and Ullman, 1979). There is also the Chomsky hierarchy for formal language recognition, which starts with a very simple automaton for the recognition of simple inputs, and ends with a Turing machine for arbitrary recursive languages (Feynman, 1996).

Next, we distinguish chaos from randomness and from complexity, but will see that there can be overlap between chaos and complexity. It's important to make the distinction because complexity is sometimes confused with randomness in the literature.

11.2.4 Chaos vs randomness vs complexity

Ideas of computational complexity have arisen within physics from the standpoints of both nonlinear dynamics[2] and statistical physics.[3] A deterministic dynamical system cannot generate truly random numbers. Deterministic chaos, which we will simply call chaos, is pseudorandomness of bounded trajectories generated via positive Liapunov exponents. The origin of pseudorandomness always lies in an algorithm. In deterministic chaos the algorithm is discovered by digitizing the underlying dynamical system and initial conditions in an integer base of arithmetic. This is not at all the same as truncating power series solutions of differential equations for computation and then using floating point arithmetic. In contrast, randomness, for example white noise or a Wiener process, is not algorithmically generated in an sde.

---

[2] See Fredkin and Toffoli (1982) for computation with billiard balls.
[3] Idealized models of neural networks are based on the Hopfield model (Hopfield, 1994; Hopfield and Tank, 1986).

Complexity is not explained by either deterministic chaos or randomness, but is a phenomenon that is distinct from either.

Deterministic dynamics generating chaotic behavior is approximated by easily predictable regular behavior over very short time scales, whereas random behavior is always unpredictable at even the shortest observable time scales. The same can be said of complexity generated by a deterministic dynamical system: over short enough time scales all deterministic systems, including chaotic and complex ones, are trivially predictable. Stochastic processes, in contrast, are unpredictable even over the shortest time scales.

Scaling is sometimes claimed to describe complexity, but scaling is an idea of simplicity: scaling is the notion that phenomena at shorter length scales look statistically the same, when magnified and rescaled, as do phenomena at larger length scales. In other words: no surprises occur as we look at smaller and smaller length scales. In this sense, the Mandelbrot set is an example of simplicity. So is the invariant set of the logistic map in the chaotic regime, where a generating partition that asymptotically obeys multifractal scaling has been discovered. Where does complexity occur in deterministic dynamics?

Edward Fredkin and Tommaso Toffoli showed in 1982 that billiard balls with reflectors (a chaotic system) can be used to compute reversibly, demonstrating that a Newtonian system is capable of behavior equivalent to a Turing machine. The difficulty in trying to use this machine in practice stems from the fact that the system is also chaotic: positive Liapunov exponents magnify small errors very rapidly. In fact, billiard balls have been proven by Ya. G. Sinai to be mixing, giving us an example of a Newtonian system that is rigorously statistical mechanical. In 1993 Moore constructed simple deterministic maps that are equivalent to Turing machines.[4] In these dynamical systems there are no scaling laws, no symbolic dynamics, no way of inferring the future in advance, even statistically. Instead of scaling laws that tell us how the system behaves at different length scales, there may be surprises at all scales. In such a system, the only way to know the future is to choose an initial condition, compute the trajectory and see what falls out. Given the initial condition, even the statistics generated by a complex system cannot be known in advance. In contrast, the statistics generated by a chaotic dynamical system with a generating partition[5] can be completely understood and classified according to classes of initial conditions. Likewise, there is no mystery in principle about which statistical distribution is generated by typical sdes.

---

[4] See Siegelmann (1995) for a connection with the Hopfield model.
[5] A generating partition is a natural, unique coarsegraining of phase space generated by the dynamical system. For chaotic one-dimensional maps, the generating partition, if it exists, is discovered via backward iteration of the (always multi-valued) map.

However, the element of complexity can perhaps be combined with stochastic dynamics as well.

Complexity within the chaotic regime is unstable due to positive Liapunov exponents, making the systems unreliable for building machines. Therefore the emphasis in the literature on the appearance of complexity at the transition to chaos.

11.2.5 Complexity at the border of chaos

In equilibrium statistical physics, universal scaling exponents arise at order-disorder transitions. For example, the transition from normal, viscous flow to superfluid flow is characterized by scaling exponents that belong to the same universality class as those for other physical systems with the same symmetry and dimension, like the planar Heisenberg ferromagnet on a three-dimensional lattice. The scaling exponents describing the vanishing of the order parameter at the critical point, the divergence of the susceptibility, and the behavior of other singular thermodynamic quantities are called critical exponents.

A related form of scaling exponent universality has also been discovered for dynamical systems at the transition to chaos where the systems under consideration are far from thermal equilibrium (Feigenbaum, 1988). For example, every map in the universality class of iterated maps defined by the logistic map generates the same scaling exponents at the transition to chaos. The same is true for the circle map universality class. This kind of universality is formally analogous to universal scaling that occurs at a second-order phase transition in equilibrium statistical physics.

It is known that limited computational capability can appear in deterministic dynamical systems at the borderline of chaos, where universal classes of scaling exponents also occur. At the transition to chaos the logistic map defines an automaton that can be programmed to do simple arithmetic (Crutchfield and Young, 1990). It is also known that the sandpile model, at criticality, has nontrivial computational capability (Moore and Nilsson, 1999). Both of these systems produce scaling laws and are examples of computational capability arising at the borderline of chaos, although the scaling exponents do not characterize the computational capability generated by the dynamics. Moore showed that simple-looking one- and two-dimensional maps can generate Turing machine behavior, and speculated that the Liapunov exponents vanish asymptotically as the number of iterations goes to infinity, which would represent the borderline of chaos (Moore, 1990, 1991; Koiran and Moore, 1999).

There is interest within statistical physics in self-organized criticality (SOC), which is the idea of a far-from-equilibrium system where the control parameter is not tuned but instead dynamically adjusts itself to the borderline of chaos (Bak *et al.*, 1987, 1988). The approach to a critical point can be modeled simply (Melby *et al.*, 2000). The logistic map, for example, could adjust to criticality without external tuning if the control parameter would obey a law of motion $D_m = D_c - a^m(D_c - D_{m-1})$ with $-1 < a < 1$ and $m = 1, 2, \ldots$, for example, where D_c is the critical value. One can also try to model self-adjustment of the control parameter via feedback from the map. However, identifying real physical dynamical systems with self-organized behavior seems nontrivial, in spite of claims that such systems should be ubiquitous in nature.

Certain scaling laws have been presented in the literature as signaling evidence for SOC, but a few scaling laws are not an adequate empirical prescription: scaling alone does not tell us that we are at a critical point, and we cannot expect critical exponents to be universal except at a critical point. SOC is an attempt to realize the notion that complexity occurs at the borderline of chaos, but scaling behavior is not complex. Moore and Nilsson (1999) have discussed the computational complexity of sandpile models. Israeli and Goldenfeld (2004, 2006) have offered a different idea of coarse-grained complexity, and have argued that there are predictable aspects. These analyses represent the frontier of complexity research and leave us without any intuitive feeling for complexity.

11.3 Replication, mutations, and reliability

We will now concentrate on "surprises," which Moore (1990) has suggested are the essence of complexity. Surprises may also describe the changes in market sentiment that lead to booms and busts. But first, some thoughts that point in the direction of surprises from computer theory and biology.

From the standpoint of physics, complex systems can do unusual things. One is self-replication. Von Neumann (1970a), who invented the first example of an abstract self-replicating automaton, also offered the following rough definition of complexity: a system is simple when it is easier to describe mathematically than to build (chaos in the solar system, for example). A system is called complex if it is easier to build or produce it than to describe it mathematically, as in the case of DNA leading to an embryo. Von Neumann's original model of a self-replicating automaton with 32 states was simplified to a two-state system by McCullough and Pitts (Minsky,

1967). The model was later generalized to finite temperatures by Hopfield (1994) and became the basis for simple neural network models in statistical physics.

Both bacteria and viruses can replicate themselves under the right conditions, but we cannot know in advance the entirely new form that a virulent bacterium might take after mutation. There, we do have not the probabilities for different possible forms for the bacteria, as in the tosses of a die. We have instead the possibility of an entirely new form, something unexpected, occurring via mutation during the time evolution of the dynamics. The result of fertilizing an egg with a sperm is another example of complexity. The essence of complexity is unpredictability in the form of "surprises" during the time evolution of the underlying dynamics. Scaling, attractors, and symbolic dynamics cannot be used to characterize complexity. From the standpoint of surprises as opposed to cataloging probabilities for a set of known, mutually exclusive alternatives, we can also see scientific progress as an example of "mutations" that may represent an underlying complex dynamical process: one cannot know in advance which new scientific discoveries will appear, nor what new technologies and also economies they may give birth to. Clearly, the economists' idea of equilibrium is completely useless for understanding inventiveness and economic growth.

Game theory, particularly Nash equilibrium ideas, is used by mainstream economic theorists (Gibbons, 1992) and has had very strong influence on the legal profession at high levels of operation (Posner, 2000). Nash equilibria have been identified as neo-classical, which partly explains the popularity of that idea (Mirowski, 2002). In econophysics, following the inventive economist Brian Arthur, the Minority Game has been extensively studied and solved by the Fribourg school (Challet *et al.*, 2005). Von Neumann first introduced the idea of game theory into economics, but later abandoned game theory as "the answer" in favor of studying automata.

Standard economic theory emphasizes optimization, whereas biological systems are apparently redundant rather than optimally efficient (von Neumann, 1970b).[6] This pits the idea of efficiency/performance against reliability, as we now illustrate. A racing motor, a sequential digital computer, and a thoroughbred horse are examples of finely tuned, highly organized machines. One small problem, one wire disconnected in a motor's ignition system, and the whole system fails. Such a system is very efficient but failure-prone. A typical biological system, in contrast, is very redundant and inefficient

---

[6] For a systematic discussion of the ideas used in von Neumann's paper, see the text by Brown and Vranesic (2000).

but has invaluable advantages. It can lose some parts, a few synapses, an arm, an eye, or some teeth, and still may function at some reduced and even acceptable level of performance, depending on circumstances. Or, in some cases, the system may even survive and function on a sophisticated level, like bacteria that are extremely adaptable to disasters like nuclear fallout. A one-legged runner is of little use, but an accountant, theorist, or writer can perform his work with no legs, both in principle and in practice. The loss of a few synapses does not destroy the brain, but the loss of a few wires incapacitates a PC, Mac, or sequential mainframe computer. Of interest in this context is von Neumann's paper on the synthesis of reliable organisms from unreliable components. Biological systems are redundant, regenerative, and have error-correcting ability. Summarizing, in the biological realm the ability to correct errors is essential for survival, and the acquisition of perfect information by living beings is impossible (see Leff and Rex (1990) for a collection of discussions of the physical limitations on the acquisition of information-as-knowledge). In economic theory we do not even have a systematic theory of correcting misinformation about markets. Instead, economics texts still feed students the standard neo-classical equilibrium line of perfect information acquisition and Pareto efficiency.[7]

In the name of control and efficiency, humanly invented organizations like firms, government, and the military create hierarchies. In the extreme case of a pure top-down hierarchy, where information and decisions flow only in one direction, downward into increasingly many branches on the organizational tree, a mistake is never corrected. Since organizations are rarely error-free, a top-down hierarchy with little or no upward feedback, one where the supposedly "higher-level automata" tend not to recognize (either ignore or don't permit) messages sent from below, can easily lead to disaster. In other words, error-correction and redundance may be important for survival. Examples of dangerous efficiency in our age of terrorism is the concentration of a very large fraction of the US's refining capacity along the Houston Ship Channel, the concentration of financial markets in New York, and the concentration of government in a few buildings in Washington, DC.

Adami (2002) and Wilke and Adami (2002) have offered ideas about evolution of complexity in digital simulations, and Velupillai (2005a) has written extensively on computability in economics, but we have no empirically based theory of either economic or biological behavior, so the existing exercises on complexity are too far removed from the phenomena we want to

---

[7] Imperfect information is discussed neo-classically, using expected utility, in the theory called "asymmetric information" by Stiglitz and Weiss (1992) and by Ackerlof (1984).

understand. In principle, complexity might emerge via bifurcations in simple models. This leads us to the notions discussed next.

11.4 Emergence and self-organization

Molecular biology is apparently largely about complexity at the cellular and molecular (DNA-protein) level. For example, the thick, impressive, and heavy text by Alberts *et al.* (2002) is an encyclopedia of cell biology, but displays no equations. Again, with no equations as an aid, Weinberg (1999) describes the five or six independent mutations required to produce a metastasizing tumor. All these impressive biological phenomena may remind us more of the results of a complicated computer program than of a dynamical system, and have all been discovered reductively by standard isolation of cause and effect in controlled, repeatable experiments. We might learn something about complexity "physically" were we able to introduce some useful equations into Alberts *et al.* The Nobel Prize winning physicist-turned-biophysicist Ivar Giæver (1999) has remarked on the difference between biology and physics texts: "Either they are right or we are right, and if we are right then we should put some equations in that text."

Many economists and econophysicists would like to use biological analogies in economics, but the stumbling block is the complete absence of a dynamical systems description of biological evolution. Instead of simple equations, we have simple objects (genes) that behave like symbols in a complicated computer program. Complex adaptable mathematical models notwithstanding, there exists no mathematical description of evolution that is empirically correct at the macroscopic or microscopic level. Schrödinger (1944), following the track initiated by Mendel[8] that eventually led to the identification of the molecular structure of DNA and the genetic code, explained quite clearly why evolution can only be understood mutation by mutation at the molecular level of genes. Mendelism (Olby, 1985; Bowler, 1989) provides us with a falsifiable example of Darwinism at the cellular level, the only precise definition of biological evolution, there being no falsifiable model of Darwinism at the macroscopic level. That is, we can understand how a cell mutates to a new form, but we do not have a picture of how a fish evolves into a bird. That is not to say that it hasn't happened, only that we don't have a model that helps us to imagine the details, which must be grounded in complicated cellular interactions that are not yet understood. Weinberg (1999) suggests that our lack of understanding of cellular networks

---

[8] Mendel studied physics before he studied peas.

also limits our understanding of cancer, where studying cellular interactions empirically will be required in order to understand how certain genes are turned on or off.

The terms "emergence" and "self-organization" are not precisely defined; they mean different things to different people. What can writers on the subject have in mind other than symmetry-breaking and pattern formation at a bifurcation in nonlinear dynamics, when they claim that a system self-organizes.[9] Some researchers who study complex models mathematically expect to discover new, "emergent" dynamics for complex systems, but so far no one has produced an empirically relevant or even theoretically clear example. See Lee (1998) for a readable account of some of the usual ideas of self-organization and emergence. Crutchfield and Young (1990), Crutchfield (1994),[10] and others have emphasized the interesting idea of nontrivial computational capability appearing/emerging in a dynamical system due to bifurcations. This doesn't present us with any new dynamics; it's simply about computational capability appearing within already existing dynamics at a bifurcation to chaos or beyond. Crutchfield assumes a generating partition and symbolic dynamics, but Moore has shown that we have to give up that idea for dynamics with Turing-equivalent complexity. Another weakness in Crutchfield is the restriction of noise to stationary processes. If we would apply that proposed method of discovery to Galilean and Keplerian orbits, then we would discover only trivial automata reflecting orbits of periods zero and one. Newton did considerably better, and we've done better in finance theory, so there must be more to the story. One can argue: the scheme wasn't invented to discover equations of motion, it was invented as an attempt to botanize complexity. In that case, can the program be applied to teach us something new and unexpected about empirical data? Why doesn't someone try to apply it to market data? Crutchfield's scheme is in any case far more specific than the program proposed by Mirowski (2002) in a related vein.

Given the prevailing confusion over "emergence," I offer an observation to try to clarify one point: whatever length and time scales one studies, one first needs to discover approximately *invariant* objects before one can hope to discover any possible new dynamic.[11] The "emergent dynamics," if such

---

[9] Hermann Haken (1983), at the Landau–Ginzburg level of nonequilibrium statistical physics, provided examples of bifurcations to pattern formation via symmetry-breaking. Too many subsequent writers have used "self-organized' as if the term would be self-explanatory, even when there is no apparent symmetry-breaking.
[10] My discussion is contrary to the philosophical expectations expressed, especially in part I, of Crutchfield's 1994 paper.
[11] For example, a cluster, like suburbanization in a city (Lee, 2004), is not an example of an approximately invariant object, because the cluster changes significantly on the length and time scale that we want to study it.

dynamics can be discovered, will be the dynamics of those objects. Now, what many complexity theorists hope and expect is that new dynamical laws beyond physics will somehow emerge statistically-observationally at larger length and time scales, laws that cannot be derived systematically from phenomena at smaller length scales. A good example is that many Darwinists would like to be able to ignore physics and chemistry altogether and try to understand biological evolution macroscopically, independently of the mass of details of genetics, which have emerged from controlled experiments and data analysis.

But consider the case of cell biology where the emergent invariant objects are genes. Genes constitute a four-letter alphabet used to make three-letter words. From the perspective of quantum physics, genes and the genetic code are clear examples of emergent phenomena. With the genetic code, we arrive at the basis for computational complexity in biology. Both DNA and RNA are known to have nontrivial computational capability (Bennett, 1982; Lipton, 1995; Adelman, 1994). One can think of the genes as "emergent" objects on long, helical molecules, DNA and RNA. But just because genes and the code of life have emerged on a long one-dimensional tape, we do not yet know any corresponding new dynamical equations that describe genetics, cell biology, or cancer. So far, one can only use quantum or classical mechanics, or chemical kinetics, in various different approximations to try to calculate some aspects of cell biology.

The point is that the *emergence of invariant objects does not imply the appearance of new laws of motion.* Apparently, invariant objects can emerge without the existence of any simple new dynamics to describe those objects. Genes obey simple rules and form four-letter words but that, taken alone, doesn't tell us much about the consequences of genetics, which reflect the most important possible example in nature of computational complexity: the evolution from molecules to cells and human life.

At a more fundamental level, genes obey the laws of quantum mechanics in a heat bath, with nontrivial intermolecular interactions. I emphasize that Schrödinger (1944) has already explained why we should not expect to discover statistically based laws that would describe evolution at the macroscale. So I am not enthusiastic about the expectation that new "emergent" laws of motion will be discovered by playing around with nonempirically inspired computer models like "complex adaptable systems." I think that we can only have hope of some success in economics, as in chemistry, cell biology, and finance, by following the traditional Galilean path and sticking close to the data. For example, we can thank inventive reductionist methods for the known ways of controlling or retarding cancer, once it develops.

11.4 Emergence and self-organization

Thinking of examples of emergence in physics, at the Newtonian level, mass and charge are invariant. The same objects are invariant in quantum theory, which obeys exactly the same local space-time invariance principles as does the Newtonian mechanics, and gives rise to the same global conservation laws. We do not yet understand how Newtonian mechanics "emerges" from quantum mechanics in a self-consistent mathematical way (quantum phase coherence cannot be destroyed in a linear theory). Similarly, we do not understand why genes should behave like elements of a classical computer, while the DNA molecule requires quantum mechanics for its formation and description. Quantum phase coherence must be destroyed in order that a Newtonian description, or classical statistical mechanics, becomes valid as a mathematical limit as Planck's constant vanishes. One can make arguments about the destruction of phase coherence via external noise in the heat bath defined by the environment, but this path only begs the question. This incompleteness in our theoretical understanding does not reduce our confidence in either classical or quantum mechanics, because all known observations of the motions of masses and charges are described correctly to within reasonable or high decimal precision at the length scales where each theory applies. One point of mesoscopic physics is to study the no-man's-land between the quantum and classical limits.

The creation of new markets depends on new inventions and their exploitation for profit. Mathematical invention has been described psychologically by Hadamard (1945). Conventional ideas of psychology (behavioral, etc.) completely fail to describe the solitary mental act of invention, whether in mathematical discovery or as in the invention of the gasoline engine or the digital computer. Every breakthrough that leads to a new invention is a "surprise" that emerges from within the system (the system includes human brains and human actions) that was not foreseen. I now suggest a simple-minded analogy between biology and economics. A completely new product is based on an invention. The creation of a successful new market, based on a new product, is partly analogous to an epidemic: the disease spreads seemingly uncontrollably at first, and then eventually meets limited or negative growth. The simplest mathematical model of creation that I can think of would be described mathematically by the growth and branching of a complete or incomplete tree (binary, ternary, ...), where new branches (inventions or breakthroughs) appear suddenly without warning. This is not like a search tree in a known computer program. Growth of any kind is a form of instability, and mathematical trees reflecting instability do appear in nature, as in the turbulent eddy cascade. But in the case of turbulence the element of surprise is missing in the dynamics.

One can also make nonempirically based mathematical and even nonmathematical models, and assert that if we assume this and that, then we expect that such and such will happen. That sort of modeling activity is not necessarily completely vacuous, because new socio-economic expectations can be made into reality by acting strongly enough on wishes or expectations: a model can be enforced or legislated, for example. Both communism (implemented via bloody dictatorships) and globalization via deregulation and privatization (implemented via legislation, big financial transfers, and supragovernmental[12] edict) provide examples. In any case, models based on real market statistics can be useful for confronting ideologues with the known constraints imposed by reality. Instead of market equilibrium, we see that instability and surprises occur with increased frequency under deregulation.

---

[12] Examples of powerful supragovernmental organizations are the IMF, the World Bank, the World Trade Organization, and the European Union. One might try to argue roughly that the US Federal Reserve Bank has had a somewhat comparable influence.

References

Ackerlof, G. A. 1984. *An Economic Theorist's Book of Tales*. Cambridge: Cambridge University Press.
Adami, C. 2002. *Bioessays* **24**, 1085.
Adelman, L. M. 1994. *Science* **266**, 1021.
Adler, W. M. 2001. *Mollie's Job: A Story of Life and Work on the Global Assembly Line*. New York: Scribner.
Alberts, B. *et al.* 2002. *Molecular Biology of the Cell*. New York: Garland Publishing.
Alejandro-Quinones, A. L. *et al.* 2005. *Proceedings of SPIE* **5848**, 27.
Alejandro-Quinones, A. L. *et al.* 2006. *Physica A* **363**, 383.
Arnold, L. 1992. *Stochastic Differential Equations*. Malabar, FL: Krieger.
Arnol'd, V. I. 1992. *Ordinary Differential Equations*. Translated by Cooke, R. New York: Springer.
Arrow, K. J. and Hurwicz, L. 1958. *Econometrica* **26**, 522.
Arthur, W. B. 1995. *Complexity* **1**, 20.
Bak, P., Tang, C., and Wiesenfeld, K. 1987. *Phys. Rev. Lett.* **59**, 381. 1988. *Phys. Rev. A* **38**, 364.
Bak, P., Nørrelykke, S. F., and Shubik, M. 1999. *Phys. Rev.* **E60**, 2528.
Ball, P. 2006. *Nature* **441**, 686.
Barbour, J. 1989. *Absolute or Relative Motion*, Vol. **1**. Cambridge: Cambridge University Press.
Barro, R. J. 1997. *Macroeconomics*. Cambridge, MA: MIT Press.
Baruch, B. 1957. *Baruch. My Own Story*. New York: Henry Holt & Co.
Bass, T. A. 1991. *The Predictors*. New York: Holt.
Bassler, K. E., Gunaratne, G. H., and McCauley, J. L. 2006. *Physica A* **369**, 343.
Bassler, K. E., McCauley, J. L., and Gunaratne, G. H. 2007. *Proc. Natl. Acad. Sci. USA* **104**, 17287.
Bassler, K. E., Gunaratne, G. H., and McCauley, J. L. 2008. *Int. Rev. Finan. Anal.* **17**, 767.
Baxter, M. and Rennie, A. 1995. *Financial Calculus*. Cambridge: Cambridge University Press.
Bell, D. and Kristol, I. (eds.) 1981. *The Crisis in Economic Theory*. New York: Basic Books.
Bennett, C. H. 1982. *Int. J. Theor. Phys.* **21**, 905.
Bernstein, P. L. 1992. *Capital Ideas: The Improbable Origins of Modern Wall Street*. New York: The Free Press.

Billingsley, P. 1983. *American Scientist* **71**, 392.
Black, F. 1986. *J. Finance* **3**, 529.
Black, F. 1989. *J. Portfolio Management* **4**, 1.
Black, F., Jensen, M. C., and Scholes, M. 1972. In Jensen, M. C. (ed.) *Studies in the Theory of Capital Markets*. New York: Praeger.
Black, F. and Scholes, M. 1973. *J. Political Economy* **81**, 637.
Bodie, Z. and Merton, R. C. 1998. *Finance*. Saddle River, NJ: Prentice-Hall.
Borland, L. 1998. *Phys. Rev. E.* **57**, 6634.
Borland, L. 2002. *Quant Finance* **2**, 415.
Bose, R. 1999 (Spring). The Federal Reserve Board Valuation Model. *Brown Economic Review*.
Bottazzi, G. and Secchi, A. 2003. *Review of Industrial Organization* **23**, 217. 2005. *Review of Industrial Organization* **26**, 195.
Bouchaud, J.-P. and Potters, M. 2000. *Theory of Financial Risks*. Cambridge: Cambridge University Press.
Bowler, P. J. 1989. *The Mendellian Revolution*. Baltimore: Johns Hopkins Press.
Bryce, R. and Ivins, M. 2002. *Pipe Dreams: Greed, Ego, and the Death of Enron*. New York: Public Affairs Press.
Caldarelli, G. 2007. *Scale-Free Networks: Complex Webs in Nature and Technology*. Oxford: Oxford University Press.
Challet, D., Marsili. M., and Zhang, Y.-C. 2005. *Minority Games: Interacting Agents in Financial Markets*. Oxford: Oxford University Press.
Cootner, P. 1964. *The Random Character of Stock Market Prices*. Cambridge, MA: MIT Press.
Cross, S. 1998. *All About the Foreign Exchange Markets in the United States*. New York: Federal Reserve Bank of New York.
Crutchfield, J. P. and Young, K. 1990. In Zurek, W. (ed.) *Complexity, Entropy and the Physics of Information*. Reading, MA: Addison-Wesley.
Crutchfield, J. 1994. *Physica D* **75**, 11.
Dacorogna, M. M. *et al.* 2001. *An Introduction to High Frequency Finance*. New York: Academic Press.
Davis, P. J. and Hersch, R. 2005. *Descartes' Dream: The World According to Mathematics*. New York: Dover.
Derman, E. 2004. *My Life as a Quant*. New York: Wiley.
Dickey, D. A., Jansen, D. A., and Thornton, D. L. 1991. *A Primer on Cointegration with Application to Money and Income*. St. Louis, MO: Federal Reserve Bank of St. Louis Review.
Dunbar, N. 2000. *Inventing Money, Long-Term Capital Management and the Search for Risk-Free Profits*. Wiley: New York.
Durrett, R. 1984. *Brownian Motion and Martingales in Analysis*. Belmont, CA: Wadsworth.
Durrett, R. 1996. *Stochastic Calculus*. Boca Raton, FL: CRC.
Eichengreen, B. 1996. *Globalizing Capital: A History of the International Monetary System*. Princeton, NJ: Princeton University Press.
Embrechts, P. and Maejima, M. 2002. *Self-similar Processes*. Princeton: Princeton University Press.
Engle, R. F. 1982. *Econometrica* **50**, 987.
Engle, R. F. and Granger, C. W. J. 1987. *Econometrica* **55**, 251.
Ezekiel, M. 1938. *Quart. J. Econ.* **52**, 255.
Fama, E. 1970. *J. Finance* **25**, 383.

Farmer, J. D. 1994. Market force, ecology, and evolution (preprint of the original version).
Farmer, J. D. Nov./Dec. 1999. Can Physicists Scale the Ivory Tower of Finance? *Computing in Science and Engineering*, 26.
Feigenbaum, M. J. 1988. *Nonlinearity* **1**, 577. 1988. *J. Stat. Phys.* **52**, 527.
Feynman, R. P. 1996. *Feynman Lectures on Computation*. Reading, MA: Addison-Wesley.
Fogedby, H., Bohr, T., and Jensen, H. J. 1992. *J. Stat. Phys.* **66**, 583.
Föllmer, H. 1995. In Howison, S. D., Kelly, F. P., and Wilmott, P. *Mathematical Models in Finance*. London: Chapman and Hall.
Frank, T. D. 2004. *Physica A*. **331**, 391.
Fredkin, E. & Toffoli, T. 1982. *Int. J. Theor. Phys.* **21**, 219.
Friedman, A. 1975. *Stochastic Differential Equations and Applications*. New York: Academic Press.
Friedman, M. & Friedman, R. 1990. *Free to Choose*. New York: Harcourt.
Friedrich, R., Peinke, J., and Renner, C. 2000. *Phys. Rev. Lett.* **84**, 5224.
Frisch, U. and Sornette, D. 1997. *J. de Physique I* **7**, 1155.
Gallegatti, M., Keen, S., Lux, T., and Ormerod, P. 2006. *Physica A* **370**, 1.
Gallucio, S., Caldarelli, G., Marsilli, M. and Zhang Y.-C. 1997. *Physica A* **245**, 423.
Geanakoplos, J. 1987. Arrow-Debreu model of general equilibrium. In Eatwell, J., Milgate, M., and Newman, P. (eds). *The New Palgrave: A Dictionary of Economics*. Basingstoke, Hampshire: Palgrave Macmillan.
Gibbons, R. C. 1992. *Game Theory for Applied Economists*. Princeton, NJ: Princeton University Press.
Gnedenko, B. V. 1967. *The Theory of Probability*. Translated by Seckler, B. D. New York: Chelsea.
Gnedenko, B. V. & Khinchin, A. Ya. 1962. *An Elementary Introduction to the Theory of Probability*. New York: Dover.
Gross, B. 2007a. http://themessthatgreenspanmade.blogspot.com/2007/10/you-reap-what-you-sow.html
Gross, B. 2007b. http://money.cnn.com/2007/11/27/news/newsmakers/gross_banking.fortune/
Gunaratne, G. 1990. Universality Beyond the Onset of Chaos. In Campbell, D. (ed.) *Chaos: Soviet and American Perspectives on Nonlinear Science*. New York: AIP.
Gunaratne, G. H. and McCauley, J. L. 2005. *Proceedings of SPIE 5848*, 131.
Gunaratne, G. H., McCauley, J. L., Nicol, M., and Török, A. 2005. *J. Stat. Phys.* **121**, 887.
Hadamard, J. 1945. *The Psychology of Invention in the Mathematical Field*. New York: Dover.
Hamermesh, M. 1962. *Group Theory*. Reading, MA: Addison-Wesley.
Hänggi, P. and Thomas, H. 1977. *Zeitschr. Für Physik* **B26**, 85.
Hänggi, P., Thomas, H., Grabert, H., and Talkner, P. 1978. *J. Stat. Phys.* **18**, 155.
Hansen, P. R. 2005. *Econometrics J.* **8**, 23.
Harrison, M. and Kreps, D. J. 1979. *Economic Theory* **20**, 381.
Harrison, M. and Pliska, S. 1981. *Stoc. Proc. Appl.* **11**, 215.
Holton, G. 1993. *Science and Anti-Science*. Cambridge, MA: Harvard University Press.
Hommes, C. H. 2002. *Proc. Natl. Acad. Sci. USA* **99**, Suppl. 3, 7221.
Hopcraft, J. E. and Ullman, J. D. 1979. *Introduction To Automata Theory, Languages, and Computation*. Reading, MA: Addison-Wesley.
Hopfield, J. J. 1994. *Physics Today*, **47**(2), 40.

Hopfield, J. J. & Tank, D. W. 1986. *Science* **233**, 625.
Hopkin, D. and Moss, B. 1976. *Automata*. New York: North-Holland.
Hughes, B. D., Schlessinger, M. F., and Montroll, E. 1981. *Proc. Natl. Acad. Sci. USA* **78**, 3287.
Hull, J. 1997. *Options, Futures, and Other Derivatives*. Saddle River, NJ: Prentice-Hall.
Hume, D. 1752. Of the Balance of Trade. In *Essays, Moral, Political, and Literary*, Vol. **1**. London: Longmans Green, 1898.
Israeli, N. and Goldenfeld, N. 2004. *Phys. Rev. Lett.* **92**, 74105.
Israeli, N. and Goldenfeld, N. 2006. *Phys. Rev. E* **73**, 26203.
Jacobs, J. 1995. *Cities and the Wealth of Nations*. New York: Vintage.
Janik, A. and Toulmin, S. 1998. *Wittgenstein's Wien*. Translated by Merkel, R. Himberg: Döcker.
Jensen, M. H., Johansen, A., Petroni, F., and Simonsen, I. 2004. *Physica A* **340**, 678.
Johansen, A., Ledoit, O. and Sornette, D. 2000. *Int. J. Theor. Appl. Finance* **3**, 219.
Johansen, S. 1991. *Econometrica* **59**, 1551.
Johansen, S. 2008. *Representation of Cointegrated Autoregressive Processes*, preprint.
Jorion, P. 1997. *Value at Risk: The New Benchmark for Controlling Derivatives Risk*. New York: McGraw-Hill.
Juselius, K. and MacDonald, R. 2003. *International Parity Relations between Germany and the United States: A Joint Modeling Approach*, preprint.
Kac, M. 1949. *Bull. Amer. Math. Soc.* **53**, 1002.
Kac, M. 1959a. *Probability and Related Topics in Physical Sciences*. New York: Wiley-Interscience.
Kac, M. 1959b. *Statistical Independence on Probability, Number Theory and Analysis*. Carus Math. Monograph no. 12. Rahway, NJ: Wiley.
Keen, S. 2001. *Debunking Economics: The Naked Emperor of the Social Sciences*. London: Zed Books.
Keen, S. 2009. *Physica A*, preprint.
Keynes, J. M. 1936. *The General Theory of Employment, Interest and Money*. New York: Harcourt, Brace and Co.
Kindleberger, C. P. 1996. *Manias, Panics, and Crashes, A History of Financial Crises*. New York: Wiley.
Koiran, P. and Moore, C. 1999. Closed-form analytic maps in one and two dimensions can simulate universal Turing machines. In *Theoretical Computer Science*, Special Issue on Real Numbers, 217.
Krugman, P. 2000. *The Return of Depression Economics*. New York: Norton.
Kubo, R., Toda, M., and Hashitsume, N. 1978. *Statistical Physics II: Nonequilibrium Statistical Mechanics*. Berlin: Springer-Verlag.
Kuersteiner, G. 2002. *MIT lecture notes on cointegration*.
Kydland, F. E. and Prescott, E. G. 1990. Business Cycles: Real Facts or Monetary Myth. *Federal Reserve Bank of Minneapolis Quarterly Review* **14**, 3.
Laloux, L., Cizeau, P., Bouchaud, J.-P., and Potters, M. 1999. *Phys. Rev. Lett.* **83**, 1467.
Lee, Y. *et al.* 1998. *Phys. Rev. Lett.* **81**, 3275.
Leff, H. S. and Rex, A. F. 1990. *Maxwell's Demon, Entropy, Information, Computing*. Princeton, NJ: Princeton University Press.
Lewis, M. 1989. *Liar's Poker*. New York: Penguin.
Lia, B. 1998. *The Society of Muslim Brothers in Egypt*. Reading: Ithaca Press.
Lillo, F. and Mantegna, R. N. 2000. *Int. J. Theor. Appl. Finance* **3**, 405.

Lillo, F. and Mantegna, R. N. 2001. *Physica A* **299**, 161.
Lipton, R. J. 1995. *Science* **268**, 542.
Lo, A. W., Mamaysky, H., and Wang, J. 2000. *J. Finance* **55**, 1705.
Lucas, R. E. 1972. *J. Economic Theory* **4**, 103.
Lucas, R. E., Jr., and Sargent, T. J. (eds.) 1982. *Rational Expectations and Econometric Practice*, New York: HarperCollins Publishers.
Luoma, J. R. 2002 (December). Water for Profit. In *Mother Jones*, 34.
Mainardi, F., Raberto, M., Gorenflo, R., and Scalas, E. 2000. *Physica A* **287**, 468.
Malkiel, B. 1996. *A Random Walk Down Wall Street*, 6th edition. New York: Norton.
Mandelbrot, B. 1966. *J. Business* **39**, 242.
Mandelbrot, B. and van Ness, J. W. 1968. *SIAM Rev.* **10**, 422.
Mankiw, N. G. 2000. *Principles of Macroeconomics*. Mason, OH: South-Western College Publishing.
Mantegna, R. E. and Stanley, H. E. 1995. *Nature* **376**, 46.
Mantegna, R. E. and Stanley, H. E. 1996. *Nature* **383**, 587.
Mantegna, R. and Stanley, H. E. 2000. *An Introduction to Econophysics*. Cambridge: Cambridge University Press.
Martin-Löf, P. 1966. *Inf. Control* **9**, 602.
Masum, H. and Zhang, Y.-C. 2004. Manifesto for the Reputation Society. *First Monday* **9**(7).
McCallum, B. T. 1989. *Monetary Economics, Theory and Policy*. New York: MacMillan.
McCann, C. R., Jr. 1994. *Probability Foundations of Economic Theory*. London: Routledge.
McCauley, J. L. 1993. *Chaos, Dynamics and Fractals: An Algorithmic Approach to Deterministic Chaos*. Cambridge: Cambridge University Press.
McCauley, J. L. 1997a. *Classical Mechanics: Flows, Transformations, Integrability and Chaos*. Cambridge: Cambridge University Press.
McCauley, J. L. 2004. *Dynamics of Markets: Econophysics and Finance*, 1st edition. Cambridge: Cambridge University Press.
McCauley, J. L. 2006. Response to Worrying Trends in Econophysics. *Physica A* **369**, 343.
McCauley, J. L. 2007. *Physica A* **382**, 445.
McCauley, J. L. 2008a. *Int. Rev. Finan. Anal.* **17**, 820.
McCauley, J. L. 2008b. *Ito Processes with Finitely Many States of Memory*, preprint.
McCauley, J. L. 2008c. *Physica A* **387**, 5518.
McCauley, J. L. and Gunaratne, G. H. G. 2003. *Physica A* **329**, 178.
McCauley, J. L., Gunaratne, G. H., and Bassler, K. E. 2007a. *Physica A* **379**, 1.
McCauley, J. L., Gunaratne, G. H., and Bassler, K. E. 2007b. *Physica A* **380**, 351.
McCauley, J. L., Bassler, K. E., and Gunaratne, G. H. 2007c. *Physica A* **387**, 302.
McCauley, J. L., Bassler, K. E., and Gunaratne, G. H. 2009. Is integration I(d) applicable to observed economics and finance time series? *Int. Rev. Finan. Anal.*
MacKay, C. 1980. *Extraordinary Popular Delusions and the Madness of Crowds*. New York: Harmony Books.
McKean, H. P. 2000. *Stochastic Integrals*. Providence, RI: AMS Chelsea.
Maslov, S. 2000. Simple model of a limit order-driven market. *Physica A* **278**, 571.
Di Matteo, T., Aste, T., and Dacorogna, M. M. 2003. *Physica A* **324**, 183.
Melby, P., Kaidel, J., Weber, N., and Hübler, A. 2000. *Phys. Rev. Lett.* **84**, 5991.
Mehrling, P. 2005. *Fischer Black and the Revolutionary Idea of Finance*. New York: Wiley.

Miller, M. H. 1988. *J. Econ. Perspectives* **2**, 99.
Millman, G. J. 1995. *The Vandals' Crown*. New York: The Free Press.
Minsky, M. L. 1967. *Computation: Finite and Infinite Machines*. New York: Prentice-Hall.
Mirowski, P. 1989. *More Heat than Light. Economics as Social Physics, Physics as Nature's Economics*. Cambridge: Cambridge University Press.
Mirowski, P. 2002. *Machine Dreams*. Cambridge: Cambridge University Press.
Modigliani, F. and Miller, M. 1958. *Amer. Econ. Rev.* **XLVIII**, 261.
Moore, C. 1990. *Phys. Rev, Lett.* **64**, 2354.
Moore, C. 1991. *Nonlinearity* **4**, 199 and 727.
Moore, C. and Nilsson, M. 1999. *J. Stat. Phys.* **96**, 205.
Morris, C. R. 2008. *The Trillion Dollar Meltdown*. New York: Public Affairs.
Murray, M. P. 1994. *The American Statistician* **48**, 37.
Muth, J. F. 1961. *Econometrica* **29**, 315.
Muth, J. F. 1982. In Lucas, R. E., Jr. and Sargent, T. J. (eds.) *Rational Expectations and Econometric Practice*. New York: HarperCollins Publishers.
Nakahara, M. 1990. *Geometry, Topology and Physics*. Bristol: IOP.
Nakamura, L. I. 2000 (July/August). Economics and the New Economy: the Invisible Hand Meets Creative Destruction. In *Federal Reserve Bank of Philadelphia Business Review* 15.
Newman, J. N. 1977. *Marine Hydrodynamics*. Cambridge, MA: MIT Press.
Niven, I. 1956. *Irrational Numbers*. Carus Math Monograph no. 11. Rahway, NJ: Wiley.
Olby, R. 1985. *Origins of Mendelism*. Chicago: University of Chicago.
Ormerod, P. 1994. *The Death of Economics*. London: Faber & Faber.
Osborne, M. F. M. 1964. In Cootner, P. *The Random Character of Stock Market Prices*. Cambridge, MA: MIT Press.
Osborne, M. F. M. 1977. *The Stock Market and Finance from a Physicist's Viewpoint*. Minneapolis, MN: Crossgar.
Plerou, V. et al. 1999. *Phys. Rev. Lett.* **83**, 1471.
Politi, M. and Scalas, E. 2008. *Physica A* **387**, 2025.
Poundstone, W. 1992. *Prisoner's Dilemma*. New York: Anchor.
Radner, R. 1968. *Econometrica* **36**, 31.
Roehner, B. M. 2001. *Hidden Collective Factors in Speculative Trading: A Study in Analytical Economics*. New York: Springer-Verlag.
Rubenstein, M. and Leland, H. 1981. *Financial Analysts Journal* **37**, 63.
Saari, D. 1995. *Notices of the AMS* **42**, 222.
Sargent, T. J. 1986. *Rational Expectations and Inflation*. New York: Harper.
Sargent, T. J. 1987. *Macroeconomic Theory*, New York: Academic Press.
Sargent, T. J. and Wallace, N. J. 1976. *Monetary Economics* **2**, 169.
Scalas, E., Gorenflo, R., and Mainardi, F. 2000. *Physica A* **284**, 376.
Scarf, H. 1960. *Int. Econ. Rev.* **1**, 157.
Schrödinger, E. 1994. *What is Life?* Cambridge: Cambridge University Press.
Schulten, K. 1999. www.ks.uiuc.edu/Services/Class/PHYS498/LectureNotes.html
Sharpe, W. F. 1964. *J. Finance* **19**, 425.
Shiller, R. J. 1999. *Market Volatility*. Cambridge, MA: MIT Press.
Siegelmann, H. T. 1995. *Science* **268**, 545.
Skjeltorp, J. A. 2000. *Physica A* **283**, 486.
Smith, A. 2000. *The Wealth of Nations*. New York: Modern Library.
Sneddon, I. N. 1957. *Elements of Partial Differential Equations*. New York: McGraw-Hill.

Solomon, S. and Levy, M. 2003. *Quantitative Finance* **3**, C12.
Sonnenschein, H. 1973. *Econometrica* **40**, 569. 1973. *J. Economic Theory* **6**, 345.
Sornette, D. 1998. *Physica A* **256**, 251.
Soros, G. 1994. *The Alchemy of Finance: Reading the Mind of the Market*. New York: Wiley.
Soros, G. 1998. *The Crisis of Global Capitalism*. New York: Little, Brown & Co.
Soros, G. 2008. *The New Paradigm for Financial Markets: The Credit Crisis of 2008 and What It Means*. New York: Public Affairs.
Steele, J. M. 2000. *Stochastic Calculus and Financial Applications*. New York: Springer-Verlag.
Stiglitz, J. E. 2002. *Globalization and its Discontents*. New York: Norton.
Stiglitz, J. E. and Weiss, A. 1992. *Oxford Economic Papers* **44**, 694.
Stratonovich, R. L. 1963. *Topics in the Theory of Random Noise*. Translated by Silverman, R. A. New York: Gordon & Breach.
Turing, A. M. 1936. *Proc. London Math. Soc.* **42**, 230.
van Kampen, N. G. 1981. *Stochastic Processes in Physics and Chemistry*. Amsterdam: North Holland.
Varian, H. R. 1992. *Microeconomics Analysis*. New York: Norton. Velupillai, K. V. 2005a. *Computability, Complexity, and Constructivity in Economic Analysis*. Oxford: Blackwell.
Varian, H. R. 1999. *Intermediate Microeconomics*. New York: Norton.
Velupillai, K. V. 2005b. *Camb. J. Econ.* **29**, 849.
von Neumann, J. 1970a. In Burks, A. W. *Essays on Cellular Automata*. Urbana, IL: University of Illinois.
von Neumann, J. 1970b. *Probabilistic Logic and the Synthesis of Reliable Elements from Unreliable Components*. In Burks, A. W. *Essays on Cellular Automata*. Urbana, IL: University of Illinois.
Wax, N. 1954. *Selected Papers on Noise and Stochastic Processes*. New York: Dover.
Weaver, W. 1982. *Lady Luck*. New York: Dover.
Weinberg, R. A. 1999. *One Renegade Cell: How Cancer Begins*. New York: Basic Books.
Wigner, E. P. 1960. *Communications in Pure and Applied Mathematics*, **13**, 1.
Wigner, E. P. 1967. *Symmetries and Reflections*. Bloomington, IN: University of Indiana.
Wilke, C. O. and Adami, C. 2002. *Trends in Ecology and Evolution* **17**, 528.
Willes, M. H. 1981. In Bell, D. and Kristol, I. (eds.) *The Crisis in Economic Theory*. New York: Basic Books.
Wilmott, P., Howe, S., and Dewynne, J. 1995. *The Mathematics of Financial Derivatives*. Cambridge: Cambridge University Press.
Wolfram, S. 1983. *Los Alamos Science* **9**, 2.
Wolfram, S. 1984. *Physica D* **10**, 1.
Yaglom, A. M. & Yaglom, I. M. 1962. *An Introduction to the Theory of Stationary Random Functions*. Translated and edited by Silverman, R. A. Englewood Cliffs, NJ: Prentice-Hall.
Yule, G. U. 1926. *J. Royal Stat. Soc.* **89**, 1.
Zhang, Y.-C. 1999. *Physica A* **269**, 30.
Zweig, S. 1999. *Die Welt von Gestern*. Frankfurt: Taschenbuch.

Index

agents, economic 8
algorithmic complexity 247
arbitrage 80
ARCH/GARCH models 238
automata 248

backward time diffusion 77, 123, 127, 180
Bak, Per 20
banking, fractional reserve 198
biology 241, 253, 256
Black, Fischer 101
Black–Scholes 106, 120, 122
Bretton Woods Agreement 196
bubbles 195, 201
Buckley, Wm. F. 200
budget deficit 203

Calls 118
capital, capital asset pricing model 115, 124
central limit theorem 36–38
chaos, deterministic 44, 250
Chapman–Kolmogorov equation 47, 57, 79
coarsegraining 150
cointegration 226
complexity 209, 241
computable numbers and functions 246
conditional average 46
conditional density 45, 74, 127, 145, 181
conservation laws 4
correlations 48, 54, 95, 101, 109, 116, 143, 160
   long time 48, 139, 141
credit 190, 201
credit, bank multiplier rule 189
credit default swaps 187

deducing dynamics from time series 162
deficits, budget 204
deflation 192
delta hedge 122, 180
demand, aggregate 21
   excess 16, 89
density, increment 164
   n-point 46
   one point 46
   transition 45
depression, liquidity crisis 195, 208
deregulation, market instability under 1, 188, 191, 194, 197, 214
derivatives 213
Derman, Emanuel 199
detrending time series or processes 50, 66, 14
diffusion and drift coefficients 50, 64
distributions, probability 44
diversification 109
dollars, explosion of in the world 197, 201, 208

econometrics 214
econophysics 7
efficient market hypothesis 94, 98
emergence 197, 256
empirical distribution 31, 146
Enron 130
ensemble (*see* statistical ensemble)
entropy 23
equilibrium 15, 28
   wrong ideas of 91, 216
ergodicity 55, 153
eurodollars 207
experiments, repeated identical 5, 36
exponential distribution 39, 182

falsifiable 3, 82
fat tails 41, 43, 103, 137, 185
finance crises 103, 188, 201, 208
finance data analysis 160
Fokker–Planck equation 67
Foley, Duncan 199
foreign exchange (FX) markets 160, 190
fractional Brownian motion 140
Friedman, Milton 203

Galileo 3
Gamblers' Ruin 84
game, chicken 206
   fair 48
   minority 8

theory, Nash 254
ultimatum 194
Gaussian density 39
   process 55
globalization 188, 201
gold standard 1, 190
green function 127, 181
Gunaratne, Gemunu H. 178

hedge 117
holism 241
Hurst exponent 43, 133, 138

Increments 54
   martingale 54
   nonstationary 53, 96, 138, 158, 236
   stationary 54, 76, 141, 155
indifference curves 20
inflation 199, 204
   runaway German 203
integrability 173
integration I(d) 226
instability, market 167, 191
International Monetary Fund 196
invariance 5, 33
   principles 4, 81
   time translational 70
invisible hand, unreliability of the 11, 18, 93, 194
Ito calculus 58
Ito process 63
Ito product 59
Ito's lemma 61, 66
Ito's theorem 58

Jacobs, Jane 203
junk bonds 199

Keynes, John Maynard 26, 214, 226
Kolmogorov 77

law, humanly invented 5
Law of large numbers 34
Law of one price 81
lawlessness, socio-economic 5
laws of nature, mathematical 4
laws of physics 4
Levy distribution 42
Liapunov exponent 245
liquidity 103, 191, 195, 203, 208, 210
local vs. global rules 62
Long Term Capital Management 105, 201
Lucas' Laissez faire policy advice 225, 235

macroeconomics 24, 214
Mandelbrot, Benoit, 96
maps, iterated 249
mark to market accounting 131
market clearing 102, 218
markets, FX 188
Markov process 46
martingale 48, 50, 62, 77, 94, 181, 160, 234

mean square fluctuation 54, 162
memory, finite-state 49, 64, 68
   infinite 142
microeconomics 13
Modigliani–Miller 85,132
monetary policy, monetarist models
   222, 224
money and uncertainty 20, 176, 187, 190
money, fiat 209
money, nonconservation of 198
money supply (M1-M3) 189, 204, 208, 210
mortgage obligations, collateralized 186, 209

neo-classical economics 10
no-arbitrage 80
noise 50, 59, 101, 227, 231
   traders 105
noncomputable 246
nonstationary process 52

Onsager, Lars 9
optimizing behavior 6, 10, 13
option price, fair 178
options 117, 177
Ornstein–Uhlenbeck–Smoluchowski model 72
Osborne, M.F.M. 24, 90

periodicity 5
phase space 16
Popper, Karl 3
portfolio 107
portfolio insurance 120, 186
price levels, relative 236
price, most probable 98, 102, 211
price, strike 117
probability 29
put-call parity 118
puts 118

Radner, Roy 20
random variable 43
rational agent 14
rational expectations 216
recurrence 99, 101, 218, 221
reductionism 241
redundance 253
regression models 214, 221, 228
regulations, effect of 193, 195, 212
reversible trading, approximately 103
risk 107, 176, 187
risk free asset 108, 187
risk premium 115

Sergeant, T.J. 222
scaling 133, 145
scaling exponents 133
scaling Ito processes 135, 165, 236
Scarf, H. 19
scatter 162
self-fulfilling expectations 190, 194, 260
self-organization 156

self-similar Ito processes (see scaling Ito processes)
shadow banking 203
sliding window, time average 169
Smith, Adam 11, 194
Soros, George, 105, 204
speculation, stabilizing 192
stagflation 199
stationary market, how to beat 193
   process 52, 70, 153
statistical ensemble 30, 36, 148
   equilibrium 72, 75, 192
   independence 46, 146
Stochastic calculus 57
   differential equations 63, 173
   differentials 58
   process 43
stretched exponential density 41
Student-t density 41, 137
stylized facts 3, 168
stylized facts, spurious 168
supply and demand 15, 18, 24
symmetry 4, 81

theory, socially constructed 216
tick time 237
time averages 153, 170

time series analysis 148
Thomas, Harry 49
topological equivalence 175
trade deficit 203
Tragedy of the Commons 194
transformations, coordinate 33, 173
Tshebychev's Theorem 35, 154
Turing, A. 5
Turing machine 248

unstable stochastic process 65
utility 13

value 98, 102, 211
   wrong ideas of 82
variance 36
Velupillai, Kumaraswamy 3
volatility 167, 238
Von Neumann 248, 253

Wallace, N. 206, 222
Walras' law 19
Wiener process 57
Wigner, Eugene 3, 4
World bank 202
World Trade Organization 202